Lecture Notes in Mathematics

Edited by A. Dold and B. Eckmann

977

T. Parthasarathy

On Global Univalence Theorems

Springer-Verlag
Berlin Heidelberg New York 1983

Author

T. Parthasarathy
Indian Statistical Institute, Delhi Centre
7, S.J.S. Sansanwal Marg., New Delhi 110016, India

AMS Subject Classifications (1980): 26-02, 26B10, 90-02, 90A14, 90C30

ISBN 3-540-11988-4 Springer-Verlag Berlin Heidelberg New York
ISBN 0-387-11988-4 Springer-Verlag New York Heidelberg Berlin

Printing and binding: Beltz Offsetdruck, Hemsbach/Bergstr.
2146/3140-543210

PREFACE

This volume of lecture notes contains results on global univalent mappings. Some of the material of this volume had been given as seminar talks at the Department of Mathematics, University of Kansas, Lawrence during 1978-79 and at the Indian Statistical Institute, Delhi Centre during 1979-80.

Even though the classical local inverse function theorem is well-known, Gale-Nikaido's global univalent results obtained in (1965) are not known to many mathematicians that I have sampled. Recently some significant contributions have been made in this area notably by Garcia-Zangwill (1979), Mas-Colell (1979) and Scarf-Hirsch-Chilnisky (1980). Global univalent results are as important as local univalent results and as such I thoughtit is worthwhile to make these results well-known to the mathematical community at large. Also I believe that there are very many interesting open problems which are worth solving in this branch of Mathematics. I have also included a number of applications from different disciplines like Differential Equations, Mathematical Economics, Mathematical Programming, Statistics etc. Some of the results have appeared only in Journals and we are bringing them to-gether in one place.

These notes contain some new results. For example Proposition 2, Theorem 4 in Chapter II, Theorem 4, Theorem 5 in Chapter III, Theorem 2" in Chapter V, Theorem 8 in Chapter VI, Theorem 2 in Chapter VII, Theorem 9 in Chapter VIII are new results.

It is next to impossible to cover all the known results on global univalent mappings for lack of space and time. For example a notable omission could be the role played by univalent mappings whose domain is complex numbers. We have also not done enough justice to the problem when a PL-function will be a homeomorphism in view of the growing importance of such functions. We have certainly given references where an interested reader can get more information.

I am grateful to Professors : Andreu Mas-Colell, Ruben Schramm, Albrecht Dold and an anonymous referee for their several constructive suggestions on various parts of this material. I am also grateful to Professor David Gale for the example given at the end of Chapter II and Professor L. Salvadori for some useful discussion that I had with him regarding Chapter VII.

Moreover I wish to thank the Indian Statistical Institute, Delhi Centre for providing the facilities and the atmosphere necessary and conducive for such work. Finally I express my sincere thanks to Mr. V.P. Sharma for his excellent and painstaking work in typing several revisions of the manuscript, Mr. Mehar Lal who typed a preliminary version of this manuscript and Mr. A.N. Sharma who helped me in filling many symbols.

1 DECEMBER 1982

T. PARTHASARATHY
INDIAN STATISTICAL INSTITUTE
DELHI CENTRE

CONTENTS

INTRODUCTION

Let Ω be a subset of R^n and let F be a differentiable function from Ω to R^n. We are looking for nice conditions that will ensure the equation $F(x) = y$ to have at most one solution for all $y \in R^n$. In other words we want the equation $F(x) = y$ to have a unique solution for every y in the range of F.

Classical inverse function theorem says that if the Jacobian of the map does not vanish then if $F(x) = y$ has a solution x^*, then x^* is an isolated solution, that is, there is a neighbourhood of x^* which contains no other solution. In the global univalence problem, we demand x^* to be the only solution throughout Ω.

It is a fascinating fact, why the global univalence problem had not been posed or at any rate solved before Hadamard in 1906, which of course is a very late stage in the development of Analysis. It is funny and actually baffling, how much misunderstanding associated with the global univalence problem survived right into the middle of the twentieth century. A brief history of this may not be out of place here.

Paul Samuelson in (1949) gave as sufficient condition for uniqueness, that the Jacobian should not vanish and it was pointed out by A. Turing that this statement was faulty. However Paul Samuelson's economic intuition was correct and in his case all the elements of the Jacobian were essentially one-signed and this condition combined with the non-vanishing determinant, turns out to be sufficient to guarantee uniqueness in the large.

Paul Samuelson (1953) then stated that non-vanishing of the leading minors will suffice for global univalence in general. But Nikaido produced a counter example to this assertion and he went on to show that global univalence prevails in any convex region provided the Jacobian matrix is a quasi-positive definite matrix. Later, Gale proved that it is sufficient for uniqueness in any rectangular region provided the Jacobian matrix is a P-matrix, that is, every principal minor is positive. In fact this culminated in the well-known article of Gale-Nikaido (1965) which is the main source of inspiration for the present writer.

I should mention two other articles. The article of Banach-Mazur (1934) gives probably the first proof of a relevant result formulated with the demands of rigour still valid to-day. The more recent article by Palais (1959) covers a much wider area than the article of Banach and Mazur.

There are several approaches one can consider to the global univalent problem. For example the approach could be via linear inequalities, monotone functions or PL functions. Throughout we have followed more or less the approach through linear inequalities.

In most of the theorems the conditions for global univalence are very stringent and therefore often not satisfied in applications. Another problem is to verify the conditions of the theorem in practice. In general it is hard to obtain necessary and sufficient conditions for global univalence results. There is lot of room for further research in this area. Gale-Nikaido's global univalent theorem is valid even if the partial derivatives are not continuous whereas Mas-Colell's results as well as Garcia-Zangwill's results demand the partial derivatives to be continuous. One of the major open problems in this area is the following: Can continuity of the derivatives in Mas-Colell's results be dispensed with (altogether or at least in part) or alternatively - are there counter examples? Another problem is the following: Is the fundamental global univalent result due to Gale-Nikaido valid in any compact convex region?

As already pointed out in some of the applications complete univalence is not warranted but in which some weaker univalence enunciations can nevertheless be made. In this connection I would like to cite at least two important papers one by Chua and Lam and the other by Schramm.

Because of the lack of a text on the global univalence and since the results are available only in articles scattered in various journals or in texts devoted to other subjects (for example Economics), I felt the need for writing this notes on global univalent mappings. In the next ten chapters with the exception of the first two chapters, various results on global univalent mappings as well as their applications are discussed. Also many examples are given and several open problems are mentioned which I believe will interest research workers.

Prerequisites needed for reading this monograph are real analysis and matrix theory. Here are a few suggestions.

[1]. W.Rudin (1976), Principles of Mathematical Analysis, Third Edition (International Student Edition) McGraw-Hill, Koyakusha Ltd.

[2]. F.R.Gantmacher (1959), The Theory of Matrices Vols. I and II, Chelsea Publishing Company, New York.

[3]. C.R. Rao (1974), Linear Statistical Inference and its Applications, Second Edition, Wiley Eastern Private Limited, New Delhi (Especially Chapter I dealing with 'Algebra of vectors and matrices').

[4]. G.S. Rogers (1980), Matrix derivatives, Marcel Dekker, New York and Basel (Actually only chapters 13 and 14 have the Jacobian and its properties as their central topic while 11 and 12 refer to the general theory).

[5]. W.Fleming (1977), Functions of several variables, Second Edition, Springer-Verlag, Heidelberg-New York.

Some knowledge of algebraic topology will be useful (especially degree theory) and we have mentioned a few references in Chapter IV.

CHAPTER I

PRELIMINARIES AND STATEMENT OF THE PROBLEM

Abstract : In this chapter we will collect some well-known results like classical inverse function theorem, domain invariance theorem etc for ready reference (without proof). We will then give the statement of the problem considered in this monograph cite a few results and make some remarks.

Classical inverse function theorem : Let F be a transformation from an open set $\Omega \subset R^n$ to R^n. We will say that F is locally univalent, if for every $x \in \Omega$ there exists a neighbourhood U_x of x such that $F|U_x$ (=F restricted to U_x) is one-one. Inverse function theorem gives a set of sufficient conditions for F to be locally univalent. We come across such problems in various situations. For example, suppose for a given y, there exists an x_o such that $F(x_o) = y$. We may like to know whether there are points x other than x_o, contained in a small neighbourhood around x_o satisfying $F(x) = y$. Classical inverse function theorem asserts that the solution is unique locally. In order to state the inverse function theorem we need the following.

Definition : A transformation F is differentiable at t_o if there exists a linear transformation L (depending on t_o) such that

$$\lim_{h \to 0} \frac{1}{||h||} [F(t_o+h)-F(t_o)-L(h)] = 0 .$$

Here $||h||$ stands for the usual vector norm. The linear transformation L is called the differential of F at t_o and is often denoted by $DF(t_o)$. Write $F = (f_1,f_2,\ldots,f_n)$ where each f_i is a real-valued function from Ω. We denote their partial derivatives as $f_i^j = \frac{\partial f_i}{\partial x_j}$.

Remark 1 : A transformation F is differentiable at t_o if and only if each of its components f_i is differentiable at t_o for $i = 1,2,\ldots,n$.

Remark 2 : If F is differentiable at t_o, then the matrix of the linear transformation L is simply the Jacobian matrix J of partial derivatives $f_i^j(t_o)$.

Definition : Call F a transformation of class q, $q \geq 0$ if each f_i is of class $C^{(q)}$. That is, for every $f_i (i = 1,2,\ldots,n)$ all the partial derivatives upto order q exist and are continuous over its domain.

We are now ready to state the (local) inverse function theorem.

Local inverse function theorem : Let F be a map of class $C^{(q)}$, $q \geq 1$ from an open set $\Omega \subset R^n$ into R^n. If the Jacobian at $t_o \in \Omega$ does not vanish, then there exists an open set $\Delta_o \subset \Omega$ containing t_o such that :

(i) $F|\Delta_o$ is one-one, that is, F restricted to Δ_o is univalent.

(ii) $F(\Delta_o)$ is an open set.

(iii) The inverse G of $F|\Delta_o$ is of class $C^{(q)}$.

(iv) $J_G(x) = (J_F(t))^{-1}$ where $F(t) = x$, $t \in \Delta_o$. Here $J_G(x)$ denotes the Jacobian matrix evaluated at x. Proof of this may be found in Fleming [17].

Remark 1 : In one dimension the situation is simpler. If F is a real-valued function with domain an open interval Ω, then F^{-1} (=inverse map of F) exists if F is strictly monotone. Also F will be strictly monotone if $F'(t) \neq 0$ for all $t \in \Omega$, and in fact $G'(x) = \frac{1}{F'(t)}$ where $x = F(t)$. In higher dimensions the Jacobian $J_F(t)$ takes the place of $F'(t)$. The situation here is more complicated. For example, the non-vanishing of the Jacobian does not guarantee that F has a (global) inverse as in the univariate case. However, if $J_F(t_o)$ does not vanish at t_o, we can find a small neighbourhood Δ_o containing t_o such that F restricted to Δ_o will have an inverse. In other words we can only assert local inverse. This is precisely part of the statement of inverse function theorem.

If one is interested in just the local univalence we have the following theorem (proof may be found in [44]).

Local univalent theorem : Let $F:\Omega \subset R^n \to R^n$ be a mapping where Ω is an open connected subset of R^n. We have the following:

(i) If F is differentiable at a point $t_o \in \Omega$ and $J_F(t_o) \neq 0$, then there is a neighbourhood U of t_o such that $F(y) = F(t_o)$, $y \in U \Rightarrow y = t_o$.

(ii) If F is continuously differentiable in a neighbourhood of an interior point t_o of Ω and $J_F(t_o) \neq 0$, then there is a neighbourhood U of t_o where F is univalent, that is, $F(y) = F(z)$, $y,z \in U \Longrightarrow y = z$.

We are now ready to state the following:

Theorem on invariance of interior points : Let $F:\Omega \to R^n$ be a differentiable map with non-vanishing Jacobian, where Ω is an open region in R^n. Then the image set $F(\Omega)$ is also an open region.

For a proof see Nikaido [44]. This result is true not only for differentiable mappings with nonvanishing Jacobians but also for homeomorphic mappings from a region of R^n into R^n. That is the content of the following classical theorem due to Brouwer.

Invariance of domain theorem: If Ω is open in R^n and $F:\Omega \to R^n$ is one-one and continuous, then $F(\Omega)$ is open and F is a homeomorphism. For a proof see [30].

Definition : A mapping $F:\Omega \to R^n$ is called a local homeomorphism if for each $t \in \Omega$, a neighbourhood of t is mapped homeomorphically by F onto a neighbourhood of F(t).

It is clear that if $F:\Omega \to R^n$ is a continuously differentiable function with non-vanishing Jacobian it follows from local inverse-function theorem or local univalent theorem that F is a local homeomorphism. We will introduce one more definition.

Definition : Let $F:\Omega \to R^n$ be a continuous mapping where Ω is an open region in R^n with the property that each $y \in F(\Omega)$ has a neighbourhood V such that each component of $F^{-1}(V)$ is mapped homeomorphically onto V by F. Then F is called a covering map and (Ω,F) is called a covering space for $F(\Omega)$. In this case, the cardinal number n of the set $F^{-1}(y)$ is the same for all $y \in F(\Omega)$. If n is a finite integer, then F is called a finite covering, or more specifically, an n-covering.

Remarks : It is well-known that every covering map $F:\Omega \to R^n$ is a homeomorphism if Ω is connected and that every homeomorphic onto function $F:\Omega \to R^n$ is a covering map and every covering map is a local homeomorphism. However the converse is not true. A local homeomorphism need not be a covering map and a covering map need not be a homeomorphic onto function. A 1-covering map is necessarily a homeomorphic onto function. The following result is well-known [48].

A theorem on covering space : Let X and Y be connected, locally pathwise connected spaces (for example $X = Y = R^n$). Furthermore suppose Y is simply connected. Then F is a homeomorphism of X onto Y if and only if (X,F) is a covering space of Y. [Here $F:X \to Y$ is a map from X to Y].

We need this result especially in chapter IV where sufficient conditions are given in order that a map F from R^n to R^n will be a homeomorphism onto R^n. For results on degree theory, we freely use from chapter VI in [48]. Other good references for degree theory are [13,59,63].

Statement of the problem : Let $F:\Omega \subset R^n \to R^n$ be a differentiable map. We want F to be globally one-one throughout Ω. What conditions should we impose on the map F and the region Ω so that F is globally one-one ?

Remark 1 : Non-vanishing of the Jacobians alone will not suffice except in the univariate case. See the example of Gale and Nikaido given in chapter III.

Remark 2 : Even in R^1 non-vanishing of the derivative is not a necessary condition for global univalence. For example $f(x) = x^3$ is globally univalent throughout R^1 whereas its derivative vanishes at $x = 0$. In general it appears difficult or hopeless to derive necessary conditions whenever global univalence prevails.

We will cite now a few typical results to give the reader some idea about this monograph.

Fundamental global univalence theorem : (Gale-Nikaido-Inada) : Let $F:\Omega \subset R^n \to R^n$ be a differentiable mapping where Ω is a rectangular region in R^n. Then F is globally univalent in Ω if either one of the following conditions holds good.

(a) $J(x)$ (= Jacobian of F at x) is a P-matrix for every $x \in \Omega$.

(b) $J(x)$ is an N-matrix and the partial derivatives are continuous for all $x \in \Omega$.

A global univalent theorem in R^3 [Parthasarathy] : Let F be a differentiable map from a rectangular region $\Omega \subset R^3$ to R^3 with its Jacobian J having the following two properties for every $x \in \Omega$:

(a) diagonal entries are negative and off-diagonal entries are positive.

(b) Every principal minor of order 2×2 is negative.

Then F is univalent in Ω.

Plastock's theorem : Let $F:R^n \to R^n$ be a continuously differentiable map. Suppose J does not vanish at any $x \in R^n$. If

$$\int_0^\infty \inf_{||x||=t} (1/||J(x)^{-1}||)dt = \infty$$

then F is a homeomorphism of R^n onto R^n. In fact F is a diffeomorphism. (Here $||x||$ stands for the usual Euclidean distant norm and $||A|| = \sup||Au||$ for A an $n \times n$ matrix and u an n vector with norm one).

In order to state McAuley's theorem we need the following definition.

Definition : Call a continuous mapping $F:\Omega \to R^n$ light if $F^{-1}(F(x))$ is totally disconnected for each $x \in \Omega$. [Here we will assume Ω to be a unit ball]. Call F open if for each U open in Ω, $F(U)$ is open relative to $F(\Omega)$. Denote by S_F the set of points $x \in \Omega$ such that F is not locally one-one at x.

McAuley's Theorem : Suppose that F is a light open mapping of a unit ball Ω in R^n onto another unit ball B in R^n such that (1) $F^{-1}F(\partial\Omega) = \partial\Omega$ (2) $F(\partial\Omega) = \partial B$ (3) $F|S_F$ is one-one (4) for each component C of $B-S_F$ there is a nonempty V in C open relative to B such that $F|F^{-1}(V)$ is one-one. Then F is a homeomorphism.

Scarf's conjecture : Let $F:\Omega \subset R^n \to R^n$ be continuously differentiable on a compact rectangle Ω with $\det J(x) > 0$ for every $x \in \Omega$. Further suppose $J(x)$ is a P-matrix for every $x \in \partial\Omega$ (= boundary of Ω). Then F is one-one throughout Ω.

This conjecture was proved by three different set of researchers Garcia-Zangwill, Mas-Colell and Scarf et al. This result is an significant generalization of Gale-Nikaido's theorem.

Schramm's theorem : Let Ω be an x-simple domain in the (x,y)-plane, ℓ its boundary. Let $F = (f,g) : \bar{\Omega} \to R^2$ be a differentiable map, α the minimum and β the maximum of f on ℓ. Suppose the Jacobian of F is an NVL matrix for each $z \in \Omega$ and for each $u \in (\alpha,\beta)$, suppose at most two points $z \in \ell$ satisfy $f(z) = u$. Then F restricted to $\bar{\Omega} \setminus (A(\alpha) \cup A(\beta))$ is univalent where $A(u) = \{z : z \in \bar{\Omega}$ and $f(z) = u\}$.

Remark 1 : Results obtained so far on global univalence are not complete and we have mentioned several interesting open problems throughout the monograph. For example it is not known whether Gale-Nikaido's result holds good in any compact convex regions. In chapter VIII and IX we have given various applications of univalent results in other areas like differential equations, Economics, Mathematical programming, Algebra etc.

Remark 2 : All the theorems cited above with the exception of McAuley's theorem depend on the choice of a fixed coordinate system. This is so because we place conditions on the Jacobian matrix. Though one may argue that this may not be the most natural approach to the problem under consideration, the present writer feels that this method yields useful results in many problems that arise in practice. See Chapter VII and Chapter IX in this connection. Also in some special cases the matrix conditions turn out to be necessary as well - see for example theorem 1 and theorem 6 in chapter VIII. Also, the present writer feels that it is not difficult to check these matrix conditions in a given problem.

CHAPTER II

P-MATRICES AND N-MATRICES

Abstract : In this chapter we will give a geometric characterization of P-matrices. We will give some properties of N-matrices. These facts we need later to prove global univalence results due to Gale, Nikaido and Inada. We will also see the interrelation between P-matrices and positive quasi-definite matrices. Finally we examine the question whether P-property holds good under multiplication (sum) of two P-matrices - this kind of result is useful in determining when the composition $F \circ G$ (sum, $F+G$) of two univalent functions is univalent.

Let A be an $n \times n$ matrix with entries real numbers. We will not consider matrices with complex entries. If A is a symmetric matrix then A is positive definite if the associated quadratic form $x'Ax > 0$, for any x different from 0. Here prime denotes the transpose of the vector x. It is well known that a symmetric matrix A is positive definite if and only if every principal minor of A is positive. Suppose we drop the symmetric assumption from A. In such situations can we prove similar results? In other words, suppose A has the following property, namely $x'Ax > 0$ for every $x \neq 0$. They can we assert that every principal minor of A is positive? Another interesting question is to characterize matrices whose principal minors are positive. Next we will answer these questions.

Characterization of P-matrices : We will start with a few definitions. Let A be a not necessarily symmetric real $n \times n$ matrix.

Definition : Call A a P-matrix if every principal minor of A is positive.

Definition : Call A a positive quasi-definite matrix if $x'Ax > 0$ for every $x \neq 0$.

Definition : Call A an N-matrix if every principal minor of A is negative. Further N-matrices are divided into two categories:

(i) An N-matrix is said to be of the first category if A has at least one positive element.

(ii) An N-matrix is said to be of the second category if every element of A is non-positive.

Definition : Call A a Leontief-type matrix if the off-diagonal entries are non-positive.

We will make a few quick remarks.

Remark 1 : Every positive quasi-definite matrix is necessarily a P-matrix (we will give a proof of this fact after characterizing the class of P-matrices). But the converse is not necessarily true as the following example shows. Let $A = \begin{bmatrix} 1 & 2 \\ 0 & 1 \end{bmatrix}$. Then $(Au,u) = u_1^2+2u_1u_2+u_2^2$ (where $u = (u_1,u_2)$) and $(Au,u) = 0$ whenever $u_1 = -u_2$. Thus A is a P-matrix but not positive quasi-definite. Also observe that A is a positive quasi-definite matrix if and only if $\left(\frac{A+A'}{2} \right)$ is a positive definite matrix. In this example $\left(\frac{A+A'}{2} \right) = \begin{bmatrix} 1 & 1 \\ 1 & 1 \end{bmatrix}$ is a singular matrix.

Remark 2 : The following example shows that every positive quasi-definite matrix need not be positive definite. Let $A = \begin{bmatrix} 2 & 2 \\ 3 & 8 \end{bmatrix}$. Clearly $(Au,u) = u_1^2+5u_1u_2+8u_2^2 > 0$ for any $u \neq 0$. Hence A is positive quasi-definite but not a positive definite matrix as A is not symmetric.

Remark 3 : First category N-matrices share some properties in common with P-matrices as we shall see below. However there is a nice characterization for symmetric second category N-matrices. In order to do that we need the following definition. Call a matrix A, merely positive definite if (i) there exists some vector x such that $x'Ax < 0$ and (ii) whenever $x'Ax < 0$, this will imply $Ax \leq 0$ or $Ax \geq 0$-in other words Ax is onesigned. The result then is the following. If A is a symmetric N-matrix of the second kind then A is merely positive definite. Furthermore A has exactly one (simple) negative eigenvalue. Proofs of these results may be found in Rao [62]. We are now ready to prove some results on P-matrices.

Theorem 1 : Let A be a P-matrix or an N-matrix of the first category. Then the system of linear inequalities

$$Ax \leq 0 \quad \text{and}$$

$$x \geq 0$$

has only the trivial solution $x = 0$.

Game theoretic interpretation of theorem 1 : Theorem 1 says that the minimax value of the matrix game A (as well as the minimax value of every principal submatrix C of A) is positive, provided A is a P-matrix or an N-matrix of the first kind. This can be seen as follows. Suppose von Neumann value of the matrix game is less than or equal to zero. (We will assume minimizer chooses rows and maximizer chooses columns). We have a probability vector y for the minimizer such that $y'A \leq 0$ or $A'y \leq 0$ (prime denotes transpose). If A is a P-matrix or an N-matrix so is A'. Thus we have got a nontrivial non-negative vector y satisfying $A'y \leq 0$ which contradicts theorem 1 and consequently value of A must be positive. It is also clear that value of A' is positive. See [49, 54] for details regarding game theory and [61] for results relating game theory and M-matrices. We follow the proof as given in

Nikaido [44].

Proof of Theorem 1 : First we will prove when A is a P-matrix. We will use induction principle. For n = 1, clearly Theorem 1, is true. So assume theorem 1 for n = k, prove that it holds good for n = k+1 (Here n refers to the order of the square matrix A). Let $x \geq 0$. That is,

$$a_{11}x_1 + a_{12}x_2 + \cdots + a_{1,k+1}\, x_{k+1} \leq 0$$
$$\cdots\cdots\cdots\cdots\cdots\cdots\cdots\cdots$$
$$a_{k+1,1}x_1 + a_{k+1,2}x_2 + \cdots + a_{k+1,k+1}x_{k+1} \leq 0$$

Since $a_{ii} > 0$, we can increase if necessary x_i such that one of the inequality becomes an equality. We will assume without loss of generality

$$a_{11}x_1 + a_{12}x_2 + \cdots + a_{1,k+1}\, x_{k+1} = 0$$

$$a_{21}x_1 + a_{22}x_2 + \cdots + a_{2,k+1}\, x_{k+1} < 0$$
$$\cdots\cdots\cdots\cdots\cdots\cdots\cdots\cdots$$
$$a_{k+1,1}x_1 + a_{k+1,2}x_2 + \cdots + a_{k+1,k+1}x_{k+1} \leq 0 .$$

Using the first equality, one can eliminate x_1 from the other inequalities. The resulting inequalities can be written as

$$a^*_{22}\, x_2 + \cdots + a^*_{2,k+1}\, x_{k+1} \leq 0$$

$$a^*_{k+1,1}x_2 + \cdots + a^*_{k+1,k+1}\, x_{k+1} \leq 0$$

where $a^*_{ij} = \dfrac{a_{11}}{a_{i1}} \times a_{ij} - a_{1j}$ where $i,j = 2,3,\ldots,k+1$. Plainly the matrix $C = (a^*_{ij})$ is a P-matrix of order k. Hence by induction hypothesis, $x_i = 0 \ \forall\ i \geq 2$. Substituting this in the first equality, $a_{11}x_1 = 0$. But $a_{11} > 0$ and hence $x_1 = 0$. This terminates the proof of theorem 1 when A is a P-matrix.

Now assume A is an N-matrix of first category. Clearly order of an N-matrix of first category should be at least 2 x 2 . Suppose $A = \begin{bmatrix} a & b \\ c & d \end{bmatrix}$ where $a < 0$, $d < 0$, and $ad-bc < 0$. This means b and c should be of the same sign. Since A is of first category, $b > 0$ and $c > 0$. Consequently $A^{-1} \geq 0$. If $Ax \leq 0$ then $A^{-1}Ax = x \leq 0$. But $x \geq 0$ by hypothesis, therefore $x = 0$. This proves the theorem when n = 2. As before assume the result for n = k where $k \geq 2$ and prove it holds good for n = k+1. As A has at least one positive element, we can imitate the proof verbatim given for P-matrices till we get the matrix $C = (a^*_{ij})i,j = 2,3,\ldots,k+1$. Observe that $\det A = a_{11} \det C$. Since $a_{11} < 0$, $\det A = a_{11} \det C < 0$, it follows that $\det C > 0$. In fact one can check that C is a P-matrix. Hence it follows from the first part of the proof $x_i = 0 \ \forall\ i \geq 2$. Since $a_{11} \neq 0, x_1 = 0$ from the first equality. This terminates the proof of theorem 1 for N-matrices of first kind.

Remark 1 : Geometrically, theorem 1 says the following: Any non-trivial non-negative vector cannot be mapped to a vector in the negative orthant when A is a P-matrix or an N-matrix of the first kind.

Remark 2 : Theorem 1 is valid for any matrix A which has non-negative inverse - that is $A^{-1} \geq 0$. Characterization results are available in the literature for such class of matrices. In particular if A is a Leontief type matrix then $A^{-1} \geq 0$ if and only if there exists some $x \geq 0$ such that $Ax > 0$.

Remark 3 : A result on linear inequalities asserts the following [See 18, pp. 49]. For any given matrix D not necessarily a square matrix exactly one of the following alternatives holds. Either the inequalities $x'D \leq 0$ has a semipositive solution or the inequality $Dy > 0$ has a non-negative solution.

In view of this result on linear inequalities, conclusion of theorem 1 can be viewed as follows: For any matrix A, suppose the system $Ax \leq 0$, $x \geq 0$ has only a trivial solution. This statement is equivalent to the fact that A has a left poverse, that is there exist non-negative matrices N,M such that $NA = I+M$. This observation is due to Charnes et. al. [8]. Another feature of the result on linear inequalities is the following, which says that von Neumann value of a P-matrix game is positive.

Theorem 2 : Suppose A is a P-matrix or an N-matrix of first category. Then there exists a positive vector $y_o \geq 0$ such that $Ay_o > 0$.

Proof : From theorem 1, the matrix $D = (A,-I)$ has no semi-positive solution x with $x'D \leq 0$ and consequently from the above remark it follows that $Dy = (A,-I)y > 0$ for some $y \geq 0$. Here $y = (y_1, y_2, \ldots, y_n, y_{n+1}, y_{n+2}, \ldots, y_{2n})$. Define $y_o = (y_1, y_2, \ldots y_n)$ Then clearly $Ay_o > z \geq 0$ where $z = (y_{n+1}, y_{n+2}, \ldots, y_{2n})$. This terminates the proof of theorem 2.

Corollary 1 : Let $S_n = \{x : x \geq 0, \sum_{i=1}^{n} x_i^2 = 1\}$. Let A be a P-matrix or or an N-matrix of first category. Then there exists an $\alpha > 0$ such that $\max_{1 \leq i \leq n} (Ax)_i \geq \alpha$ for every $x \in S_n$.

Proof : From theorem 1, for every $x \in S_n$ it follows that Ax has at least one coordinate strictly positive. Since $\max_{1 \leq i \leq n} (Ax)_i$ is continuous in x and S_n is compact, $\min_{x \in S_n} \max_{1 \leq i \leq n} (Ax)_i = \max_{1 \leq i \leq n} (Ax^o)_i$ for some $x^o \in S_n$. Set $\alpha = \max_{1 \leq i \leq n} (Ax^o)_i$. This α will satisfy the requirements of the corollary and the proof is complete.

In order to give a characterization theorem for P-matrix, we introduce the following definition:

Definition : Let A be an $n \times n$ matrix. The matrix A is said to reverse the sign of a vector $x \in R^n$ if $(Ax)_i\, x_i \leq 0$ for all $i = 1,2,\ldots,n$. (Here as usual $(Ax)_i$ denotes the i-th coordinate of the vector Ax).

Theorem 3 : Let A be an $n \times n$ real matrix. Then A is a P-matrix if and only if A does not reverse the sign of any vector except the zero vector.

Proof (Necessity) : Let A be a P-matrix. Suppose A reverses the sign of a non-zero vector x. If $x \geq 0$ then $(Ax)_i\, x_i \leq 0$ for all i, will imply $Ax \leq 0$. In other words we have a semipositive solution to the system of inequalities $Ax \leq 0$ and $x \geq 0$, but this contradicts theorem 1. Let $L = \{j : x_j < 0\}$ and let D be the diagonal matrix with diagonal entries +1 or -1 in the j-th column according as $j \notin L$ or $j \in L$. Trivially $Dx \geq 0$ and Dx is not a zero vector for $x \neq 0$. Also $D^{-1} A\, D$ is a P-matrix and it reverses the sign of a non-negative vector Dx which is not possible as shown before. This completes the necessity part. We will now prove, (Sufficiency). Let A be any $n \times n$ matrix which does not reverse the sign of any non-zero vector. Note that every principal sub-matrix C of A also cannot reverse the sign of any non-zero vector. As such it is enough to show that det A is positive. Suppose det $A \leq 0$, then (since det A is equal to the product of the eigenvalues) there will exist a real eigenvalue $\lambda \leq 0$. Let $Ax = \lambda x$ where x is the corresponding eigenvector. Clearly $(Ax)_i x_i = \lambda x_i^2 \leq 0$. In other words A reverses the sign of a non-zero vector which is a contradiction. This terminates the proof of theorem 3.

One obvious consequence of theorem 3 is the following corollary.

Corollary 2 : Suppose A is a non-singular matrix. Then A is a P-matrix if and only if A^{-1} (= inverse of A) is a P-matrix.

Proof : Suppose A is a P-matrix and A^{-1} is not a P-matrix. Then there exists $x \neq 0$, such that A^{-1} reverses the sign of x. This will imply A will reverse the sign of $y = A^{-1}x \neq 0$ and hence a contradiction (via theorem 3) to the supposition that A is a P-matrix. This completes the proof of corollary 2.

We will now identify several well known classes of P-matrices.

Proposition 1 : (i) Any positive quasi-definite matrix is a P-matrix (ii) Any matrix having a positive dominant diagonal is a P-matrix and (iii) Any matrix that satisfies Stolper-Samuelson condition (A matrix A is said to satisfy Stolper-Samuelson condition if $A \geq 0$, A is non-singular and A^{-1} is of Leontief type) is a P-matrix.

Proof : (i) Let A be positive quasi-definite. If A is not a P-matrix there exists a vector $x \neq 0$ such that $(Ax)_i\, x_i \leq 0$. [This is a consequence of theorem 3]. Hence $x'Ax \leq 0$ which contradicts our hypothesis regarding A. Therefore every positive quasi-definite matrix is a P-matrix. We will now prove (iii). Proof of (ii) will be taken up last. Let A be a matrix with $A \geq 0$, A non-singular and A^{-1} Leontief type. We will first prove that the von Neumann value associated with A^{-1} is positive. Suppose value of A^{-1} is less than or equal to zero. This will imply the existence of a probability vector y such that $yA^{-1} \leq 0$. [Here we assume that the rows are chosen by the minimizer and the columns by the maximizer]. Since $A \geq 0$, $yA^{-1}A \leq 0$ or $y \leq 0$ which is a contradiction. Hence value of $A^{-1} > 0$. That is, there exists a probability vector x_0 such that $A^{-1} x_0 > 0$. Since A^{-1} has off-diagonal entries non-positive it follows that A^{-1} has positive dominant diagonal property. [We say that a matrix B of order $n \times n$ has positive dominant diagonal property if there exists a strictly positive vector $d = (d_1, d_2, \ldots, d_n)$ where each $d_i > 0$ such that $b_{ii} d_i > \sum_{\substack{j=1 \\ j \neq i}}^{n} |b_{ij}| d_j$ for every $i = 1,2,\ldots,n$]. For the moment assume (ii) holds good. Then A^{-1} is a P-matrix and the corollary after theorem 3 tells us that A is a P-matrix. Thus we need to prove only (ii) to complete the proof of proposition 1.

Let A be a matrix having a positive dominant diagonal. That is, $d_1>0, d_2>0, \ldots, d_n > 0$ such that $a_{ii} d_i > \sum_{\substack{j=1 \\ j \neq i}}^{n} |a_{ij}| d_j$. Clearly $a_{ii} > 0$ for $i = 1,2,\ldots,n$. First we will prove that $\det A \neq 0$. Then we will show det A is positive. Suppose $\sum_{j=1}^{n} a_{ij} x_j = 0$ for $i = 1,2,\ldots,n$ where $x = (x_1, x_2, \ldots, x_n)$ is a non-zero vector. Define

$$\theta = \max_{1 \leq i \leq n} \frac{|x_i|}{d_i} = \frac{|x_k|}{d_k} \text{ say .}$$

Then $\theta \geq 0$. In case $\theta = 0$, then x will be zero vector which will contradict the fact that x is a non-zero vector. Suppose $\theta > 0$. We have

$$a_{kk} x_k = - \sum_{\substack{j=1 \\ j \neq k}}^{n} a_{kj} x_j . \text{ Then } |a_{kk} x_k| \leq \Big| \sum_{\substack{j=1 \\ j \neq k}}^{n} a_{kj} x_j \Big| \leq \theta \sum_{j \neq k} |a_{kj}| d_j < \theta\, a_{kk} d_k .$$

Since $a_{kk} > 0$, $|x_k| < \theta d_k$. That is $\theta < \theta$ which is impossible. This shows that $\det A \neq 0$. Now it will be shown that $\det A > 0$. If A has a positive dominant diagonal then $A + \rho I$ also has a positive dominant diagonal for every $\rho \geq 0$. Note that $\det (A+\rho I)$ is a continuous function in ρ and that $\det (A + \rho I) \to +\infty$ as $\rho \to +\infty$. If $\det A \leq 0$ this will imply $\det (A + \rho_o I) = 0$ for some $\rho_o > 0$ which is clearly impossible. Hence $\det A > 0$. If A has a positive dominant diagonal, every principal submatrix C of A also has the same property and consequently A is a P-matrix. This terminates the proof of the proposition. [Thus we are fortunate enough to identify several classes of P-matrices. We are able to characterize only symmetric N-matrices of the second kind. The situation is far from complete in the case of non-symmetric N-matrices]. However one can prove the following elementary result for N-matrices of order 3×3 of the first kind [53].

Proposition 2 : Let A be an N-matrix of the first kind with order 3×3. Then A will contain exactly four positive elements and five negative elements.

Proof : Since A is an N-matrix of the first kind and since A is of order 3×3, it is clear that the number of positive elements is either 2 or 4 or 6. Suppose A has 2 positive elements. Then A will be of the form :

$$A = \begin{bmatrix} - & + & - \\ + & - & - \\ - & - & - \end{bmatrix} .$$

Clearly det $A > 0$ which is impossible. Similar contradiction will be reached if the off-diagonal entries are positive. Thus it follows that A contains exactly four positive elements and five negative elements. This terminates the proof of proposition 2.

Remark 1 : If A is a positive definite or a quasi-positive definite matrix and if Q is a non-singular matrix then it follows that Q'AQ, is a positive definite or a quasi-positive definite matrix. Such a result is not valid in general for P-matrices (though it is true if Q is a diagonal matrix with all entries +1 or -1. This fact is used during the course of our proof of Theorem 3). Let $A = \begin{bmatrix} 1 & -4 \\ 0 & 2 \end{bmatrix}$ and $Q = \begin{bmatrix} 4 & 1 \\ 0 & 1 \end{bmatrix}$. Then $Q'AQ = \begin{bmatrix} 16 & -12 \\ 4 & -1 \end{bmatrix}$. Clearly A is a P-matrix while Q'AQ is not a P-matrix.

Remark 2 : If A is an N-matrix of the first kind and if Q is a diagonal matrix with all entries +1 or -1 then Q'AQ is an N-matrix but it may be of second kind.

Open problem : Let A be a positive definite matrix. Can one find a diagonal matrix D with diagonal entries strictly positive such that DA-I is a (weak) N-matrix? [We call a matrix B, a weak N-matrix if every principal minor of B is non-positive].

If the answer to this question is in the affirmative, this will settle an old conjecture of Paul Levy regarding the infinite divisibility of multivariate gamma distribution - See Paranjape [51].

A final remark on N-matrices of the first kind: If A is an n x n, N-matrix of the first kind, then every principal minor of order less than or equal to n-1 of A^{-1} is positive. (Note that det $A^{-1} < 0$ since det $A < 0$).

Is AB a P-matrix or N-matrix when A and B are P-matrices or N-matrices ? In this section we are interested in examining whether the product of two P-matrices (or N-matrices) will be a P-matrix (or N-matrix). In this connection we have the following [53].

<u>Theorem 4</u> : (i) There exist two P-matrices (N-matrices) A and B such that AB is not a P-matrix (N-matrix).

(ii) If A and B are 2 × 2 Leontief type P-matrices (N-matrices of the same kind) then AB is a P-matrix.

(iii) If A and B are 3 × 3 Leontief type P-matrices then AB is a P-matrix but AB need not be of the Leontief type.

(iv) If A and B are n × n Leontief type B-matrices and AB is also of Leontief type then AB is a P-matrix. However this result is not valid if we do not assume that AB is of Leontief type when $n \geq 4$ [see (iii) above].

<u>Proof</u> : (i) $A = \begin{bmatrix} 1 & -3 \\ 0 & 1 \end{bmatrix}$, $B = \begin{bmatrix} 1 & 0 \\ 2 & 1 \end{bmatrix}$. Then $AB = \begin{bmatrix} -5 & -3 \\ 2 & 1 \end{bmatrix}$ is not a P-matrix. If we assume all the entries to be positive then clearly AB is a P-matrix if A and B are 2 × 2 P-matrices. However such a result is not true for 3 × 3 P-matrices. Let

$$\begin{bmatrix} 1 & 1 & 1 \\ 1 & 2 & 4 \\ 1 & 4 & 12 \end{bmatrix}, \quad B = \begin{bmatrix} 1 & 1 & 1 \\ 1 & 2 & 1 \\ 1 & 1 & 2 \end{bmatrix}. \quad \text{Then} \quad AB = \begin{bmatrix} 3 & 4 & 4 \\ 7 & 9 & 11 \\ 17 & 21 & 29 \end{bmatrix}$$

Clearly A and B are positive-definite matrices but AB is not a P-matrix as the principal minor $\begin{bmatrix} 3 & 4 \\ 7 & 9 \end{bmatrix}$ is negative.

(ii) Proof of this is elementary and we omit it. However a word of caution is necessary. If A,B are 2 × 2 matrices with A, an N-matrix of the first kind, B and N-matrix of the second kind then AB need not be neither a P-matrix nor an N-matrix as the following example demonstrates. Let $A = \begin{bmatrix} -1 & 2 \\ 1 & -1 \end{bmatrix}$, $B = \begin{bmatrix} -3 & -7 \\ -1 & -2 \end{bmatrix}$ then $AB = \begin{bmatrix} 1 & 3 \\ -2 & -5 \end{bmatrix}$ is neither a P-matrix nor an N-matrix.

(iii) We will prove that AB is a P-matrix provided A and B are 3 × 3 Leontief type P-matrices. Let

$$A = \begin{bmatrix} a_{11} & -a_{12} & -a_{13} \\ -a_{21} & a_{22} & -a_{23} \\ -a_{31} & -a_{32} & a_{33} \end{bmatrix}, \quad B = \begin{bmatrix} b_{11} & -b_{12} & -b_{13} \\ -b_{21} & b_{22} & -b_{23} \\ -b_{31} & -b_{32} & b_{33} \end{bmatrix}.$$

Here a_{ij} and b_{ij} are all non-negative. It is clear that $\det(AB) = \det A . \det B > 0$ and further the diagonal entries are positive in AB. We need only to check that every principal minor of order 2 is positive. Let us take the leading principal minor C of order 2 from AB. Then

$$C = \begin{bmatrix} a + a_{13}\, b_{31} & b + a_{13}\, b_{32} \\ c + a_{23}\, b_{31} & d + a_{23}\, b_{32} \end{bmatrix}$$

where

$$\begin{bmatrix} a & b \\ c & d \end{bmatrix} = \begin{bmatrix} a_{11} & -a_{12} \\ -a_{21} & a_{22} \end{bmatrix} \times \begin{bmatrix} b_{11} & -b_{21} \\ -b_{21} & b_{22} \end{bmatrix}$$

Clearly $ad - bc > 0$. Also note $b = -a_{11}\, b_{12} - a_{12}\, b_{22} \leq 0$ and $c = -a_{21}b_{11} - a_{22}b_{21} \leq 0$. Then

$$\det C = (ad-bc) - a_{13}\, b_{32}\, c - a_{23}\, b_{32}\, b + a_{23}\, b_{32}\, a + a_{13}\, b_{31}\, d$$
$$> 0.$$

Similarly one can show the other two principal minors of order 2 in AB are positive. Thus we have shown that AB is a P-matrix.

(iv) We will show that AB is a P-matrix if A,B are Leontief type P-matrices and if AB is of Leontief type. Since A and B are Leontief type P-matrices it follows that A^{-1} and B^{-1} are non-negative. Consequently $(AB)^{-1}$ is non-negative. In other words $(AB)^{-1}$ satisfies Stolper-Samuelson condition and therefore AB is a P-matrix. We will now give a counter example to show that this result is false if we drop the assumption AB is of Leontief type when $n \geq 4$. Let,

$$A = \begin{bmatrix} 1 & 0 & -1 & 0 \\ 0 & \frac{1}{8} & 0 & -1 \\ 0 & 0 & 1 & 0 \\ 0 & 0 & 0 & 1 \end{bmatrix} \quad \text{and } B = \begin{bmatrix} 1 & -1 & 0 & 0 \\ -1 & 2 & 0 & 0 \\ 0 & -4 & 1 & 0 \\ -1 & 0 & 0 & 1 \end{bmatrix}$$

Then,

$$AB = \begin{bmatrix} 1 & 3 & -1 & 0 \\ 7/8 & 2/8 & 0 & -1 \\ 0 & -4 & 0 & 0 \\ -1 & 0 & 0 & 1 \end{bmatrix}.$$

One can check that A and B are Leontief type P-matrices but AB is not of Leontief type. Also note that AB is not a P-matrix as one of the principal minors of order 2, $\begin{bmatrix} 1 & 3 \\ 7/8 & 2/8 \end{bmatrix}$ is not positive. This terminates the proof of theorem 4.

As already pointed out elsewhere, the motivation for proving theorem 4 is simply that it helps us to determine when F o G will be univalent given that F and G are differentiable and univalent in R^n. Gale-Nikaido's fundamental result asserts global univalence of a differentiable mapping F when the Jacobian of the map F is a P-matrix. Note that the Jacobian associated with F o G is equal to the product of the Jacobian matrices associated with F and G. We will prove the following result in another chapter. If F and G are differentiable maps from R^3 (or R^2) to R^3 (or R^2) and if their Jacobians are P-matrices everywhere then F o G is a P-function and consequently one-one throughout R^3 (or R^2). From theorem 4 we may be able to conclude such a result is valid in R^n for any n provided we assume that the Jacobian associated with F o G is of Leontief type.

If A and B are 2 × 2 Leontief type P-matrices then A+B is of Leontief type but need not be a P-matrix as the following simple example demonstrates. Let $A = \begin{bmatrix} \frac{1}{2} & -1 \\ 0 & \frac{1}{2} \end{bmatrix}$ and $B = \begin{bmatrix} \frac{1}{2} & 0 \\ -1 & \frac{1}{2} \end{bmatrix}$. Then $A+B = \begin{bmatrix} 1 & -1 \\ -1 & 1 \end{bmatrix}$ is not a P-matrix. Such results will be useful to determine when F+G is one-one given that F and G are one-one, differentiable functions. Observe the following result which is easy to prove. If A and B are n x n Leontief type P-matrices and if there exists a non-negative vector $x \geq 0$ such that $(A+B)x > 0$ then A+B is a Leontief type P-matrix. [Here $(A+B)x > 0$ means, every component of the vector $(A+B)x$ is positive].

It may not be out of place to mention other interesting equivalent properties of P-matrices. We will state it in the form of a theorem without proof [56].

Theorem 5 : Let A be an n × n real matrix. Then each of the following conditions is equivalent to the statement : A is a P-matrix

(i) Every real eigenvalue of each principal submatrix of A is positive.

(ii) A+D is nonsingular for each nonnegative diagonal matrix D

(iii) For each $x \neq 0$ there exists a nonnegative diagonal matrix D such that $x'ADx > 0$

(iv) A does not reverse the sign of any nontrivial vector.

(v) For each signature matrix S there exists an $x > 0$ such that $SASx > 0$.

Remark 1 : We have already shown that (iv) is equivalent to the statement that A is a P-matrix. We can use this fact to prove the equivalence of other statements. When A is of Leontief type, proof can be given using ideas from game theory-interested readers should refer to Raghavan [61]. Such a beautiful characterization is not available for N-matrices. We will close this chapter by mentioning two open problems: (1) Suppose A is a P-matrix. Does it mean A is positive stable? [Call a matrix A positive stable if the real part of each eigenvalue of A is positive]. Clearly this is so if either A is a Leontief type P-matrix or if A is a 2 × 2 P-matrix. (2) Suppose A is a positive definite matrix. Does there exist a diagonal matrix D with entries strictly positive such that DA-I is a (weak) N-matrix (that is principal minors of (DA-I) is non-positive). Again this result is true if A is a 2×2 positive definite matrix.

David Gale has communicated to the author the following example which answers the first question raised above in the negative. Let

$$A = \begin{bmatrix} \varepsilon & 1 & 0 \\ 0 & \varepsilon & 1 \\ 1 & 0 & \varepsilon \end{bmatrix}$$

Then A has eigenvalues $1+\varepsilon$, $\varepsilon-\frac{1}{2} + i\sqrt{3}/2$ with negative real parts if $\varepsilon < \frac{1}{2}$ and of course A is a P-matrix.

We also wish to add that if A is a positive definite or quasi-positive definite matrix then A is positive stable. We will indicate the proof in the quasi-positive definite case. In fact as already remarked elsewhere A is quasi-positive definite if and only if $(A+A')/2$ is a P-matrix. Let λ_m and λ_M be the minimal and maximal eigenvalues of $(A+A')/2$. Let λ be any eigenvalue of A. Then it is well-known that

$$\lambda_m \leq \text{Real part of } \lambda \leq \lambda_M \quad .$$

Since $\lambda_m > 0$, it follows that A is positive stable.

CHAPTER III

FUNDAMENTAL GLOBAL UNIVALENCE RESULTS OF GALE-NIKAIDO-INADA

Abstract: In this chapter we will derive fundamental global univalence results of Gale-Nikaido-Inada for differentiable maps in rectangular regions whose Jacobian matrices are either P-matrices or N-matrices. Using mean-value theorem of differential calculus we will obtain global univalence results for differentiable maps whose Jacobian matrices are quasi-positive definite matrices, in convex regions. This result is due to Gale and Nikaido. Finally we present two new results on univalent mappings in R^3.

Let F be a differentiable map from $\Omega \subset R^n$ to R^n. We are interested in finding suitable conditions such that F is univalent throughout Ω. Non-vanishing of the Jacobian matrices will not suffice as we shall see below. There are at least two approaches to the problem under consideration. One approach places topological assumptions on the map and the other places further conditions on the Jacobian matrices. We will study the former in the next chapter and the latter in the present chapter.

Global univalence theorems : In order to state and prove results on univalent mappings we will adopt the following notation and terminology. We closely follow Gale and Nikaido's original paper. Let Ω be a rectangular region in R^n, where $\Omega = \{x : x \in R^n,\ a_i \leq x_i \leq b_i\}$. Here a_i, b_i are real numbers where we may allow some or all of them to assume $-\infty$ or $+\infty$. Let $F:\Omega \to R^n$ be a mapping defined by $F(x) = (f_1(x), f_2(x), \ldots, f_n(x))$ where each $f_i(x)$ is a real-valued function defined on Ω. Then F is said to be differentiable in Ω just in case every f_i has a total differential $\Sigma f_{ij}(x)dx_j$ in Ω - in other words for x, $a \in \Omega$, $f_i(x) = f_i(a) + \sum_j f_{ij}(a)(x_j - a_j) + O(||x-a||)(i=1,2,\ldots,n)$, where $O(||x-a||)/||x-a|| \to 0$ as $x \to a$.

The Jacobian matrix J of the mapping F is given by $J(x) = ||f_{ij}(x)||$. Differentiability of F clearly implies continuity of F and it also implies partial differentiability of f_i's and in fact $f_{ij}(x) = \partial f_i(x)/\partial x_j$. However, one has to be careful at the boundary points. Yet the Jacobian can be defined by means of the coefficients of the total differential at those points. Now we are ready to prove a non-linear counterpart of theorem 1 of chapter II.

Theorem 1 : Let F be a differentiable mapping from Ω to R^n. Suppose the Jacobian matrix J(x) of the mapping F is a P-matrix for every $x \in \Omega$. Then for any a and x in Ω, the inequalities $F(x) \leq F(a)$ and $x \geq a$ have only one solution namely $x = a$.

Proof : We prove this theorem by induction on n, the common dimension of both the argument and image vectors $x, F(x)$. Plainly theorem is true for $n = 1$. Let

$$X = \{x : x \in \Omega \text{ and } F(x) \leq F(a),\ x \geq a\}\ .$$

Clearly $a \in X$. We are through if we show that X contains only a. With that in mind, we first prove that a is an isolated point of X. As F is differentiable, we have

$$\lim_{x \to a} \left|\left| \frac{F(x)-F(a)}{||x-a||} - J(a) \frac{x-a}{||x-a||} \right|\right| = 0.$$

Since J(a) is a P-matrix, there is a positive number $\delta > 0$ by corollary 1, chapter II such that for any $x \geq a$, some coordinate of $J(a) \frac{x-a}{||x-a||} \geq \delta > 0$. Consequently, in any neighbourhood of a, some component of $F(x)-F(a)$ must be positive for $x \geq a$ in Ω. This shows that a is an isolated point of X. Suppose $b \in X$ with $b \neq a$. Clearly $b \geq a$. Define $Y \subset X$ as follows .

$$Y = \{x : a \leq x \leq b,\ F(x) \leq F(a)\}\ .$$

Plainly Y is compact and since a is an isolated point, $Y - \{a\}$ is also compact. Let $\bar{x}$ be a minimal element of $Y - \{a\}$ in the sense that no other element $y \in Y - \{a\}$ fulfills $y \leq \bar{x}$ (Note that such an $\bar{x}$ can be obtained by minimizing the sum of the components of y when y ranges over $Y - \{a\}$). As $\bar{x} \in Y - \{a\}$, only two possibilities can occur (i) $\bar{x} > a$, (ii) $\bar{x} \geq a$ and $\bar{x}$ has some component equal to the corresponding component of a.

Case (i) : $\bar{x} > a$. Theorem 2 of chapter II ensures the existence of a vector u satisfying $u < 0$ and $J(\bar{x})\, u < 0$ (for $J(\bar{x})$ is a P-matrix). Define $x(t) = \bar{x} + tu$. As $u < 0$, $\bar{x} > a$ and Ω is a rectangular region for sufficiently small positive $t, x(t) \in \Omega$. Moreover by differentiability $\frac{F(x(t))-F(\bar{x})}{t||u||} - J(\bar{x}) \frac{u}{||u||}$ can be made as small as we please by letting t approach 0. Since $J(\bar{x})\, u < 0$ it follows for small positive $t, F(x(t)) < F(\bar{x}) \leq F(a)$ and consequently $x(t) \in Y - \{a\}$. Note that $x(t) < \bar{x}$ and this contradicts the minimality of $\bar{x}$. This rules out the possibility of case (i).

Case (ii) : Now we will apply induction principle. We will assume without loss of generality that $\bar{x}_1 = a_1$ where $\bar{x}_1$ and a_1 are the first coordinates of $\bar{x}$ and a respectively. We now define a new differentiable mapping $G : \hat{\Omega} \to R^{n-1}$ where

$$\hat{\Omega} = \{(x_2, x_3, \ldots, x_n) : (a_1, x_2, \ldots, x_n) \in \Omega\}$$

and

$$g_i(x_2, x_3, \ldots, x_n) = f_i(a_1, x_2, x_3, \ldots, x_n),\ i = 2, 3, \ldots, n$$

Jacobian matrix of G is plainly a principal submatrix of J(x) and hence it is a P-matrix . Further

$$g_i(\bar{x}_2,\bar{x}_3\ \ldots,\bar{x}_n) \leq g_i(a_2,a_3,\ldots,a_n)$$

$$\bar{x}_i \geq a_i \quad \text{for} \quad i = 2,3,\ldots,n.$$

Therefore we must have $\bar{x}_i = a_i$ for $i = 2,3,\ldots,n$ by induction hypothesis. This contradicts $\bar{x} \neq a$. Hence case (ii) is also impossible. This terminates the proof of theorem 1.

For N-matrix we have the following [44].

Theorem 2 : Let F be a continuously differentiable mapping from $\Omega \to R^n$ $(n \geq 2)$. Suppose $J(x)$ is an N-matrix of the first category for each $x \in \Omega$. Then $F(x) \leq F(a)$ and $x \geq a$ imply $x = a$.

Proof : Proof of this theorem is almost the same as that of theorem 1 except that one has to exercise caution for two reasons: (i) induction will start when $n = 2$ (ii) Althouth every principal submatrix of an N-matrix is also an N-matrix they need not be of the same category. In other words we not only have to show that the Jacobian of the new mapping G to be an N-matrix but also to be of the first category.

(i) We will now prove theorem 2 when $n = 2$. As F is continuously differentiable all the partial derivatives are continuous. Consider the Jacobian matrix $J(x)$:

$$J(x) = \begin{bmatrix} f_{11} & f_{12} \\ \\ f_{21} & f_{22} \end{bmatrix}$$

As $J(x)$ is of first category, $f_{11} < 0$, $f_{22} < 0$ and $f_{21} > 0$, $f_{12} > 0$. As all the partial derivatives are continuous, they will be of definite sign throughout $x \in \Omega$. Define $G = (f_2,f_1)$. Jacobian matrix of G is a P-matrix. Clearly $F(x) \leq F(a)$ iff $G(x) \leq G(a)$. Using theorem 1, we can complete the argument.

(ii) We will now show that the Jacobian of the map G which appears under case (ii) of theorem 1, is an N-matrix of the first kind in order to complete the induction argument of the present theorem. The rest of the proof is same as that of theorem 1. Define $\hat{J}(x_2,x_3,\ldots,x_n)$ = principal minor of $J(x)$ got by omitting the first column and the first row where $x = (a_1,x_2,\ldots,x_n)$. Clearly $\hat{J}$ is an N-matrix. Suppose every element of $\hat{J}$ is non-positive. If $f_{1j}(\bar{x}_1,\bar{x}_2,\ldots,\bar{x}_n) > 0$ for $j=2,3,\ldots,n$ then $f_{1j}(x) > 0$ for every $x \in \Omega$ - this is due to the fact that all partial derivatives should be of definite signs (as already pointed out in (i)). But since $\bar{x}_1 = a_1, \bar{x}_i \geq a_i$, $i = 2,\ldots,n$ with strict inequality for atleast one $i \geq 2$, we have $f_1(\bar{x}) > f_1(a)$ contradicting $F(x) \leq F(a)$. Therefore $f_{1j_0}(\bar{x}_1,\bar{x}_2,\ldots,\bar{x}_n) \leq 0$ for some j_o. Define a vector v which has zero everywhere except at the j_oth coordinate where it is 1. This vector would be a nontrivial solution to $J(\bar{x})v \leq 0$ which contradicts theorem 1 of chapter II. Hence $\hat{J}$ must be an N-matrix of the first kind. This completes the discussion for theorem 2.

Remark : Theorem 1 (as well as 2) can be proved under somewhat weaker restrictions on the Jacobian matrix. For example theorem 1 holds good if for every x, J(x) as well as its principal submatrices and their transpose matrices considered as matrix games have positive minimax values. We have already pointed out that every P-matrix has this property but the converse is not true. For example the matrix $\begin{bmatrix} 1 & 2 \\ 4 & 8 \end{bmatrix}$ is not a P-matrix but every game that can be constructed out of this in the manner described as above has positive value. In order to prove the main result due to Gale, Nikaido and Inada we need the following definition.

Definition : A mapping $F:\Omega \to R^n$ is a P-function if for any $x,y \in \Omega$, $x \neq y$, there is an index $k = k(x,y) \in \{1,2,\ldots,n\}$ such that $(x_k-y_k)(f_k(x)-f_k(y)) > 0$. Trivially a P-function is a 1-1 function [42].

Fundamental Global Univalence Theorem (Gale-Nikaido-Inada) : Let $F:\Omega \to R^n$ be a differentiable mapping where Ω is a rectangular region in R^n. Then F is globally univalent on Ω if either one of the following conditions holds good.

(a) J(x) is a P-matrix $\forall\ x \in \Omega$

(b) J(x) is an N-matrix and the partial derivatives are continuous for all $x \in \Omega$. In fact, F is a P-function under condition (a).

Proof : (under condition a) : We assume that J(x) is a P-matrix $\forall\ x \in \Omega$. We will show that F is a P-function, using induction principle. For n = 1, result is obvious. We will assume the result to be true for n-1 and show that it is true for n. Let $x,y \in \Omega$ be such that $x \neq y$. Suppose $(x_i-y_i)(f_i(x)-f_i(y)) \leq 0$. We will also assume that $x_i \neq y_i$ for all i = 1,2,...,n. For if $x_i = y_i$ for some i, we can construct a mapping G similar to that as in theorem 1, and then use induction hypothesis to exhibit an index k such that $(x_k-y_k)(f_k(x)-f_k(y)) > 0$. Let D be a diagonal matrix whose i-th diagonal entry is +1 or -1 according as $x_i > y_i$ or $x_i < y_i$. Let H be a mapping from $D(\Omega) \to R^n$ defined as follows: $H(z) = D \circ F \circ D(z)$ for every $z \in D(\Omega)$. Plainly $D(\Omega)$ is a rectangular region and the Jacobian matrix of the mapping H is again a P-matrix. Let $x^* = D(x)$ and $y^* = D(y)$. Since $(x_i-y_i)(f_i(x)-f_i(y)) \leq 0$ and $x_i \neq y_i$ for all i, it follows that $H(x^*) \leq H(y^*)$ where $x^* > y^*$, but this contradicts theorem 1. This terminates the proof of the first part.

Proof (under condition b) : We will now prove univalence when J(x) is an N-matrix. Suppose F(x) = F(y) and $x \neq y$. Let D be a diagonal matrix with i-th diagonal entry +1 or -1 according as $x_i \geq y_i$ or $x_i < y_i$. As in part (a), let H be a mapping from

$D(\Omega) \to R^n$ defined by $H = D \circ F \circ D$. It is clear that the Jacobian J_H of H will be an N-matrix, the partial derivatives will be continuous and they will have definite signs. Also throughout, J_H will be of the same category. If J_H is of first category, then $H(x^*) = H(y^*)$ and $x^* = D(x) \geq y^* = D(y)$ imply (from theorem 2) that $x^* = y^*$ or $x = y$ which contradicts our assumption $x \neq y$. If J_H is of second category then every member of J_H will be strictly negative. If $x^* \neq y^*$, then $H(x^*) \leq H(y^*)$, with the inequality strict in some coordinate, contradicting the fact that $H(x^*) = H(y^*)$. Hence as before $x^* = y^*$ or $x = y$ contradicting $x \neq y$. This finishes the proof of part (b) and the theorem.

Remark 1 : Continuity of the partial derivatives is crucially used when $J(x)$ is an N-matrix. It is not clear whether the continuity of the partial derivatives can be omitted from the theorem.

Remark 2 : We have assumed Ω to be a rectangular region in theorem 1,2 and the fundamental theorem. How far can one relax this assumption? Will it suffice if we assume Ω to be a convex region? The following theorem gives a partial answer, when the Jacobian is everywhere a positive quasidefinite matrix - in this case proof is also extremely simple which uses mean value theorem of differential calculus of a single variable.

Theorem 3 : Let $\Omega \subset R^n$ be a convex set and $F:\Omega \to R^n$ be a differentiable map, with its Jacobian everywhere positive (or negative) quasidefinite in Ω. Then F is univalent.

Proof : We will prove only for positive quasi-definite case as the proof for negative quasi-definite case is similar. Suppose $a,b \in \Omega$ with $a \neq b$. We will directly show that $F(a) \neq F(b)$. Let $x(t) = ta + (1-t)b = b + t(a-b) = b + th$ where $h = a-b \neq 0$ by assumption. By convexity, $x(t) \in \Omega$ for $t \in [0,1]$. Let $\phi(t) = \Sigma_i h_i(f_i(x(t)) - f_i(b))(1 \geq t \geq 0)$. Differentiating with respect to t, we have $\phi'(t) = \sum_j \sum_i h_i h_j f_{ij}(x(t))$ which is positive everywhere by positive quasidefiniteness of the Jacobian. Since $\phi(0) = 0$, $\phi(1) \neq 0$. That is $\Sigma h_i(f_i(a) - f_i(b)) \neq 0$ or $f_i(a) \neq f_i(b)$ for some i. Consequently $F(a) \neq F(b)$. This proves F is univalent.

As an application we have the following corollary.

Corollary (Noshira, 1934) : Let $g(z)$ be an analytic (complex) function of a complex variable z in a convex region Ω of the complex plane. Suppose its derivative $g'(z)$ has positive real part in Ω. Then $g(z)$ is univalent in Ω.

Proof : Let $g(z) = u(x,y) + i\,v(x,y)$ where $i^2 = -1$ and $z = x + iy$. Then $g'(z) =$

$\frac{\partial u}{\partial x} + i\frac{\partial v}{\partial x}$. Note $\frac{\partial u}{\partial x} > 0$. Now, consider the mapping $F : \Omega \to R^2$ defined by $F(x,y) = (u(x,y), v(x,y))$. Its Jacobian is given by

$$J = \begin{bmatrix} \frac{\partial u}{\partial x} & \frac{\partial u}{\partial y} \\ \frac{\partial v}{\partial x} & \frac{\partial v}{\partial y} \end{bmatrix} .$$

However from the Cauchy-Riemann equations we have

$$\frac{\partial u}{\partial x} = \frac{\partial v}{\partial y}, \quad \frac{\partial u}{\partial y} = -\frac{\partial v}{\partial x} .$$

Hence

$$\frac{J+J'}{2} = \begin{bmatrix} \frac{\partial u}{\partial x} & 0 \\ 0 & \frac{\partial u}{\partial x} \end{bmatrix} \text{ is a P-matrix for } \frac{\partial u}{\partial x} > 0.$$

This implies that J is positive quasidefinite and from theorem 3 one can conclude that F is univalent. This terminates the proof of the corollary.

The following example shows that theorem 3 may not hold good in a non-convex region. Let $g(z) = z + \frac{1}{z}$ $(z \neq 0)$. Let Ω be the common exterior portion of two circles of radius $\frac{1}{2}$ having their centres at $\frac{1}{2}$ and $-\frac{1}{2}$ respectively. One can verify that the Real part of $g'(z)$ is positive in Ω, as

$$\mathrm{Re}(g'(z)) = \frac{(|z-\frac{1}{2}|^2-\frac{1}{4})(|z+\frac{1}{2}|^2-\frac{1}{4}) + (\mathrm{Im}(z))^2}{|z|^4} .$$

However $g(i) = g(-i) = 0$ and $i, -i \in \Omega$. Theorem 3 may fail even if the region is simply connected. In fact in the above example we can take any simply connected subregion of Ω containing i and -i where theorem 3 will fail.

In the next chapter we will prove univalence results which place analytical conditions on the map F. Also we will prove an important result of More and Rheinboldt which uses results from degree theory.

<u>Univalent results in R^3</u> : We will present two univalent results for differentiable maps in R^3. One result is valid for rectangular regions and the other for any convex region in R^3 [53].

We will start with a simple example. Let $F = (f,g,h)$ where $f(x,y,z) = x^2 + z^2$, $g(x,y,z) = x^2+y^2$ and $h(x,y,z) = y^2 + z^2$. Let Ω be any convex region in the strictly positive orthant which includes points of the form $(4, \frac{1}{2}, z)$. In this example,

$$J = \begin{bmatrix} 2x & 0 & 2z \\ 2x & 2y & 0 \\ 0 & 2y & 2z \end{bmatrix} \quad \text{and} \quad \frac{J+J'}{2} = \begin{bmatrix} 2x & x & z \\ x & 2y & y \\ z & y & 2z \end{bmatrix} .$$

Observe that J is a P-matrix but $\frac{J+J'}{2}$ is not a P-matrix since the 2 x 2 leading minor is (equal to $4xy-x^2$) is negative at $x = 4, y=\frac{1}{2}$. Thus the Jacobian matrix is not positive-quasidefinite and it is not difficult to check that the map F is one-one in any convex region in the strictly positive orthant. In the example note that f does not depend on y, g does not depend on z and h does not depend on x.

Theorem 4 : Let $F = (f,g,h)$ be a differentiable map from a convex region $\Omega \subset R^3$ to R^3. Suppose f does not depend on x, g does not depend on y and h does not depend on z. Further suppose the partial derivatives f_y, f_z, g_x, g_z, h_x and h_y are positive throughout Ω. Then F is univalent over Ω.

Proof : Jacobian matrix J of F can be written as :

$$J = \begin{bmatrix} 0 & f_y & f_z \\ g_x & 0 & g_z \\ h_x & h_y & 0 \end{bmatrix}$$

Suppose $F(a) = F(b)$ with $a \neq b$. Let $a = (a_1,a_2,a_3)$ and $b = (b_1,b_2,b_3)$. We need to discuss essentially two cases (i) $a_i \geq b_i$ for $i = 1,2,3$ and (ii) $a_1 < b_1$, $a_2 \geq b_2$ and $a_3 \geq b_3$. (Other cases can be reduced to one of these two cases).

Case (i) : $a_i \geq b_i$ for $i = 1,2,3$. Note that $f(a_1,a_2,a_3) = f(b_1,a_2,a_3)$ since f is independent of the first variable. We will assume without loss of generality $a_2 > b_2$. Then, since $f_y > 0$ and $f_z > 0$, $f(a_1,a_2,a_3) = f(b_1,a_2,a_3) > f(b_1,b_2,a_3) \geq f(b_1,b_2,b_3)$. In other words $f(a_1,a_2,a_3) > f(b_1,b_2,b_3)$ which contradicts our assumption that $F(a) = F(b)$.

Case (ii) : $a_1 < b_1$, $a_2 \geq b_2$ and $a_3 \geq b_3$. First we will prove when $a_1 < b_1$, $a_2 > b_2$, $a_3 > b_3$. As in the proof of the fundamental theorem define a 3 x 3 diagonal matrix $D = \begin{bmatrix} -1 & 0 & 0 \\ 0 & 1 & 0 \\ 0 & 0 & 1 \end{bmatrix}$ and the map $H:D(\Omega) \to R^3$ when $H(u) = D \circ F \circ D(u)$. It is not difficult to check that the first row of the Jacobian matrix associated with H will have the following property. First entry will be identically zero while the second and third entries will be negative. Let $D(a) = a^*$ and $D(b) = b^*$. Then $H(a^*) = H(b^*)$ with $a^* \geq b^*$ and as in case (i) we can argue to show that $h_1(a^*) < h_1(b^*)$ (Here h_1 is the first component function of H) which leads to a contradiction. In case $a_1 < b_1$, $a_2 > b_2$, $a_3 = b_3$ then clearly $f(a_1,a_2,a_3) = f(b_1,a_2,a_3) > f(b_1,b_2,a_3) = f(b_1,b_2,b_3)$ which again contradicts our assumption that $F(a) = F(b)$. This terminates the proof of theorem 4.

Remark 1 : One interesting feature of this result as well as the next is that we have not explicitly assumed that the Jacobian is non-vanishing. Automatically this is satisfied because of our other conditions.

Remark 2 : We do not know whether a similar result is true in R^n for $n > 3$. At the present we do not have a counter example in R^4.

We are ready to state our next theorem [53].

Theorem 5 : Let F be a differentiable map from a rectangular region $\Omega \subset R^3$ to R^3 with its Jacobian J having the following two properties for every $x \in \Omega$:

(i) diagonal entries are negative and off-diagonal entries are positive.

(ii) Every principal minor of order 2 x 2 is negative. Then F is univalent in the rectangular region.

Before proving this result we would like to make the following observations. First note that the Jacobian is not an N-matrix. It can be seen as follows. Signs of the entries of J can be written out explicitly because of condition (i).

$$J = \begin{bmatrix} - & + & + \\ + & - & + \\ + & + & - \end{bmatrix} .$$

In other words -J is of Leontief type matrix. Also one can easily check by expanding through first row that determinant $J > 0$ - this is a consequence of condition (i) and (ii). Second, we need to prove theorem 1 in this situation. Proof of theorem 1 crucially depends on theorem 1 of chapter II. We first prove the following.

Lemma 1 : A be a 3 x 3 matrix with the following two properties.

(i) diagonal entries are negative and off-diagonal entries are positive.

(ii) Principal minors of order 2 x 2 are negative. Then the system of inequalities $Ax \leq 0$ and $x \geq 0$ has only the trivial solution $x = 0$.

Proof : Suppose $Ax \leq 0$, $x \geq 0$ and $x \neq 0$. Let $x = (x_1, x_2, x_3)$.

$$A = \begin{bmatrix} a_{11} & a_{12} & a_{13} \\ a_{21} & a_{22} & a_{23} \\ a_{31} & a_{32} & a_{33} \end{bmatrix} . \text{ If } x_3 = 0 \text{ then } \begin{bmatrix} a_{11} & a_{12} \\ a_{21} & a_{22} \end{bmatrix} \begin{bmatrix} x_1 \\ x_2 \end{bmatrix} \leq 0, (x_1, x_2) \neq 0$$

which is clearly impossible since $\begin{bmatrix} a_{11} & a_{12} \\ a_{21} & a_{22} \end{bmatrix}$ is an N-matrix of the first kind.

So we will assume $x_i > 0$ for $i = 1,2,3$. We have,

$$a_{11}x_1 + a_{12}x_2 + a_{13}x_3 \leq 0$$

$$a_{21}x_1 + a_{22}x_2 + a_{23}x_3 \leq 0 .$$

Since, $a_{13}x_3 > 0$ and $a_{23}x_3 > 0$, we have, $a_{11}x_1 + a_{12}x_2 \leq 0$ and $a_{21}x_1 + a_{22}x_2 \leq 0$, which is again impossible. This terminates the proof of lemma 1.

Lemma 2 : Let F be a differentiable map from a rectangular region $\Omega \subset R^3$ with its Jacobian J having the following two properties for every $x \in \Omega$: (i) diagonal entries are negative and off-diagonal entries are positive (ii) every principal minor of order 2 x 2 is negative. Then $F(x) \leq F(y)$ and $x \geq y$ imply $x = y$. Proof of lemma 2 is similar to the proof of theorem 1 (and theorem 2) and hence omitted.

Proof of theorem 5 : Let $F = (f,g,h)$. Suppose $F(a) = F(b)$ with $a \neq b$. If $a \geq b$ then from lemma 2, we infer $a = b$ which leads to a contradiction. Let $a=(a_1,a_2,a_3)$ and $b = (b_1,b_2,b_3)$. If $a_i=b_i$ for some i, we can define a new map G in R^2 as we have done in the proof of theorem 1 where the Jacobian of G will be an N-matrix of order 2 x 2 of the first kind and using this fact we will be able to show that $a = b$ which leads to a contradiction. From now onwards we will assume that $a_i \neq b_i$ for any i=1,2,3. Let $a_1 < b_1$, $a_2 > b_2$ and $a_3 > b_3$. Let $D = \begin{bmatrix} -1 & 0 & 0 \\ 0 & 1 & 0 \\ 0 & 0 & 1 \end{bmatrix}$ and $H = D \circ F \circ D$. It is easy to check that every element in the first row of the Jacobian matrix of the map H is strictly negative. Let $D(a) = a^*$, $D(b) = b^*$. Then $H(a^*) = H(b^*)$ and $a^* > b^*$. Let $H = (h_1,h_2,h_3)$. Hence $h_1(a^*) < h_1(b^*)$ and this contradicts the fact that $H(a^*) = H(b^*)$. Similar contradictions will be reached in other cases. This terminates the proof of theorem 5.

Remark : It is not known whether theorem 5 remains true in convex regions. Also it is not clear how to formulate theorem 5 in higher dimensions.

We will now give two examples. The first example demonstrates that the positivity of the leading minors will not suffice for global univalence as originally suggested by Paul Samuelson, while the second example shows that P-property (of the Jacobian matrices) is not a necessary condition for global univalence.

Example 1 : Let $F = (f,g)$ where $f(x,y) = e^{2x}-y^2+3$ and $g(x,y) = 4y\, e^{2x}-y^3$. Let $\Omega = \{(x,y): -2 \leq x,y \leq 2\}$. Then the Jacobian J of F is given by

$$J = \begin{bmatrix} 2e^{2x} & -2y \\ 8y\, e^{2x} & 4e^{2x}-3y^2 \end{bmatrix} .$$

Clearly $2e^{2x}$ and det $J = 8e^{4x} + 10y^2 e^{2x}$ are positive throughout R^2 and F is not one-one as $F(0,2) = F(0,-2) = (0,0)$. Trouble arises in this example because the function $4e^{2x}-3y^2$ that appears in the Jacobian matrix changes sign. This example is due to Gale and Nikaido [19].

Example 2 : Let $F = (f,g)$ where $f(x,y) = \frac{x^3}{3} - \frac{x}{2} - y$, $g(x,y) = x+y$. Here the Jacobian J is given by

$$J = \begin{bmatrix} x^2-\frac{1}{2} & -1 \\ 1 & 1 \end{bmatrix}$$

Plainly J is not a P-matrix when $x^2 < \frac{1}{2}$. However F is one-one throughout R^2 as the following argument shows. Let $A = [\begin{smallmatrix} 1 & 1 \\ 0 & 1 \end{smallmatrix}]$. Then $AJ = [\begin{smallmatrix} x^2+\frac{1}{2} & 0 \\ 1 & 1 \end{smallmatrix}]$ is a P-matrix. In other words the map $G = (f+g,g)$ is one-one in R^2 by Gale-Nikaido's theorem and consequently F is one-one in R^2. [Note that in the first example AJ can never be a P-matrix in Ω for any non-singular matrix A].

If theorem 1 can be proved for convex regions then Gale-Nikaido's theorem will remain valid for convex regions. The following two problems appear to be challenging open problems (1) Does theorem 1 remain true for compact convex regions? (2) Does Inada's theorem on global univalence remain valid for rectangular regions when the Jacobian matrix is an N-matrix and if we drop the assumption that the partial derivatives are continuous. In other words does theorem 2 remain valid for differentiable map F which is not necessarily of order $C^{(1)}$?

A result due to Kestelman asserts the following:

Let F be a continuously differentiable map from $\Omega \to R^n$ where Ω is an open set in R^n with $J(x)$ non-singular for all $x \in \Omega$. Let K be a compact subset of Ω with nonempty interior. Then F_K(= F restricted to K) is one-one if ∂K(= boundary of K) is connected and if $F_{\partial K}$ (= F restricted to ∂K) is one-one. For a proof of this assertion see [31]. The assertion is false if ∂K is not connected as the following example shows. Let K be the set in the complex plane defined as follows:

$$K = \{z : |z| \leq 1 \quad \text{or} \quad |z-2\pi i| \leq \tfrac{1}{2}\} .$$

Let $F(z) = e^z$. Clearly $F(0) = F(2\pi i) = 1$ and K is not connected.

In Gale-Nikaido's theorem Ω is a rectangular region - we will assume without loss of generality Ω to be a compact rectangular region. We will also assume $\Omega \subset R^n$ where $n \geq 2$. Then clearly $\partial\Omega$ is a connected set. Also suppose F is continuously differentiable with positive Jacobian throughout Ω. In view of Kestelman's result, in order F to be one-one it is enough if we check that F is one-one on the boundary $\partial\Omega$. This naturally raises the following important question: Is F univalent if we simply assume that the Jacobian is a P-matrix for only those x which belong to the boundary of Ω ? Scarf has conjectured that this question will have an affirmative answer. Indeed this conjecture has been verified recently by three sets of researchers: (i) C.Garcia and W.Zangwill (ii) G.Chichilnisky, M.Hirsch and H.Scarf and (iii) A. Mas-Colell. This will form the subject matter of a subsequent chapter. Garcia-Zangwill's proof of this conjecture depends on norm-coerciveness theorem while the proof of Mas-Colell depends on the Poincare index theorem of differential topology. It is not clear how one can use Kestelman's result directly to give an alternative proof of Scarf's conjecture.

We pose another related question: "Is F univalent when the Jacobian is an N-matrix for every $x \in \partial\Omega$? ". This is certainly true in R^2 but we do not know the answer in R^n for $n \geq 3$.

CHAPTER IV

GLOBAL HOMEOMORPHISMS BETWEEN FINITE DIMENSIONAL SPACES

Abstract : It is well-known from covering space theory that global homeomorphism problem can be reduced to finding conditions for a local homeomorphism to satisfy the line lifting property. We will show that this property is equivalent to a limiting condition (which in many cases easy to verify) which we call by L. We will use this condition L to derive several results on global homeomorphisms due to Roy Plastock. We will prove an approximation theorem due to More and Rheinboldt and this result will then be used to prove Gale-Nikaido's theorem under weaker assumptions. In the last section we will prove a result due to McAuley for light open mappings. We will end this chapter with an old conjecture of Whyburn.

Line lifting property equivalent to condition (L) : We will start with a few definitions. Let F be a map from R^n to R^n.

Definition : A continuous map F is said to be proper if $F^{-1}(K)$ is compact whenever K is compact.

Definition : Let $\Omega \subset R^n$ be open and connected. Then $F:\Omega \to R^n$ lifts lines in $F(\Omega)$ if for each line $L(t) = (1-t)y_1 + ty_2$ $(0 \leq t \leq 1)$ in $F(\Omega)$ and for every point $x_\alpha \in F^{-1}(y_1)$ there is a path $P_\alpha(t)$ such that $P_\alpha(0) = x_\alpha$ and $F(P_\alpha(t)) = L(t)$.

Remark : If F is a local homeomorphism and F lifts lines in $F(\Omega)$, then the path $P_\alpha(t)$ in the above definition is unique for each α.

Let Ω be open and connected in R^n. Let $F:\Omega \to R^n$ be continuous. We now introduce the condition (L).

Condition (L) : Whenever $P(t)$, $0 \leq t < b$, is a path satisfying $F(P(t)) = L(t)$ for $0 \leq t < b$ (where $L(t) = (1-t)y_1 + ty_2$ is any line in R^n), then there is a sequence $t_i \to b$ as $i \to \infty$ such that $\lim_{i\to\infty} P(t_i)$ exists and is in Ω.

We need the following well-known results for the proof of Plastock's theorem [55].

Lemma : Let X and Y be connected, locally pathwise connected spaces where X and $Y \subset R^n$. Furthermore let Y be simply connected. Then F is a homeomorphism of X onto Y if and only if F is a covering map of X onto Y.

For definitions regarding simply connected regions see [33].

Theorem (Hermann) : Let $\Omega \subset R^n$ be open and connected, $F:\Omega \to R^n$. In order that F is a covering map of Ω onto $F(\Omega)$ it is necessary and sufficient that (i) F is a local homeomorphism and (ii) F lifts lines in $F(\Omega)$.

For a proof of this theorem see Hermann [28] or Plastock (p. 170-171, [55]).

We now have the following result due to Plastock.

Plastock's Theorem : Let $F:\Omega \subset R^n \to R^n$ be a local homeomorphism. Then condition (L) is both necessary and sufficient for F to be a homeomorphism of Ω onto R^n.

Proof : In view of Hermann's theorem, to prove the sufficiency, it is enough if we show that F lifts lines in $F(\Omega)$. Let $L(t)$ be any line in $F(D)$ with $L(0) = y$ and let $x \in F^{-1}(y)$.

We can find an $\varepsilon > 0$ and a path $P(t)(=F^{-1}(L(t)))$, $0 \leq t < \varepsilon$ with $P(0) = x$ and $F(P(t)) = L(t)$ for $0 \leq t < \varepsilon$. Let $c(\leq 1)$ be the largest number for which $P(t)$ can be extented to a continuous path for $0 \leq t < c$ and satisfying $F(P(t)) = L(t)$, $0 \leq t < c$. Let $z = \lim_{t_i \to c} P(t_i)$ and observe that this limit exist for the map F satisfies condition (L). By continuity, $F(z) = L(c)$. Let U be a neighbourhood of z on which F is a homeomorphism. There exists N_0 such that $P(t_i) \in U$ for $i \geq N_0$. Also there exists a $\delta > 0$ and a path $Q(t)$ defined for $c-\delta<t<c+\delta$ so that $Q(t_M)=P(t_M)$ (where M is chosen so that $M \geq N_0$ and $c-\delta<t_M<c$) and $F(Q(t))=L(t)$ for $c-\delta<t<c+\delta$.

Hence $P(t)$ can be extended to a continuous path (which we continue to call $P(t)$) on $0 \leq t < c+\delta$, $P(0) = x$ and $F(P(t)) = L(t)$, $0 \leq t < c+\delta$. By the maximality of c, we conclude that $c = 1$, and hence F lifts lines. By virtue of Hermann's theorem F is a covering map from $\Omega \to F(\Omega)$. We need only show that $F(\Omega) = R^n$ in order to apply the above lemma and thus conclude that F is a homeomorphism of Ω onto R^n. So let $y \in R^n$. Choose $y_1 \in F(\Omega)$ and let $L(t) = (1-t)y_1 + ty$. If we retrace the steps of the first part of our proof, we find a path $P(t)$, $0 \leq t \leq 1$, so that $F(P(t)) = L(t)$ on $0 \leq t \leq 1$. In particular $F(P(1)) = L(1) = y$, and so $F(\Omega) = R^n$. This completes the proof of Plastock's theorem.

Remark : The same proof goes through even in infinite dimensional Banach spaces. We will now deduce Browder's theorem from this theorem.

Browder's Theorem : $F:R^n \to R^n$ is a homeomorphism if and only if F is a local homeomorphism and a closed map.

Proof : We will only check the sufficiency part. We will show that F satisfies condition (L). Suppose $P(t)$ is defined on $0 \leq t < b$ and satisfies $F(P(t)) = L(t)$

for $0 \leq t < b$. Let S = closure of $\{P(t): 0 \leq t < b\}$. Since F is a closed map F(S) is closed. Thus since $L(t) \in F(S)$, for all $t < b$, by continuity $L(b) \in F(S)$. This means for some $x \in S$, $F(x) = L(b)$. That is we can find $\{t_i\}$ such that $P(t_i) \to x$. Since t_i is a bounded sequence, without loss of generality we will assume $t_i \to t_0$. We claim $t_0 = b$. By continuity $L(t_0) = L(b)$ and hence $t_0 = b$. Thus condition (L) is satisfied. Now Browder's Theorem is a consequence of Plastock's theorem.

Remark 1 : It is easy to see that Hadamard's theorem as well as Banach-Mazur's theorem is a consequence of Browder's theorem. Their result simply states that $F: R^n \to R^n$ is a homeomorphism if and only if F is a proper map and a local homeomorphism. It is easy to see that every proper map is a closed map.

Remark 2 : If F is a local homeomorphism and if F is also closed then F is proper.

We will now give an analytic condition which will in turn imply condition (L). In order to do that we need the following. Let W(x) be a strictly positive real-valued continuous function on R^n. Let P(t) be a path in R^n of class $C^{(1)}$ on $0 \leq t \leq b$.

Definition : The arc length of P with weight W is

$$L_o^b(P) = \int_0^b W(P(t))||P'(t)||dt .$$

Here P'(t) stands for the derivative of P(t) and $||\cdot||$ stands for the usual matrix norm.

Definition : R^n is complete with respect to arc length with weight W if $L_o^b(P) < \infty$ implies $\lim_{t \to b} P(t)$ exists and is finite whenever P(t) is a $C^{(1)}$ path on $0 \leq t < b$.

Remark : The above definition is equivalent to the usual notion of R^n being complete with respect to the conformal metric induced by the tensor $ds^2 = [W(x)]^2 dx^2$. See [28]. Definition given above is meaningful even in infinite dimensional Banach spaces. We now have the following theorem due to Plastock.

Theorem 2 : Let $F: R^n \to R^n$ be a continuously differentiable map. Suppose Jacobian J of F does not vanish at any $x \in R^n$. If

$$\int_0^\infty \inf_{||x||=t} (1/||J(x)^{-1}||)dt = \infty ,$$

then F is a homeomorphism of R^n onto R^n. In fact F is a diffeomorphism (that is F is one-one onto and its inverse is also differentiable).

Proof : Define the weight function $W(x) = (1/||J(x)^{-1}||)$. Let $u(s) = \inf_{||x||=s} W(x)$.

We are given that $\int_0^\infty u(s)ds = \infty$. We will first show that R^n is complete with respect to arc length with weight W.

Let $P(t) \in C^{(1)}[0,b)$ and suppose $L_0^b(P) < \infty$. Let $0 < \delta < b$. For any partition $0 = t_0 \leq t_1 \leq \ldots \leq t_N = \delta$ of $[0,\delta]$, let $t_i \leq \bar{t}_i \leq t_{i+1}$ be that number for which $\sup_{t_i \leq t \leq t_{i+1}} ||P'(t)|| = ||P'(\bar{t}_i)||$. Here $P'(t)$ stands for the derivative of $P(t)$.

From the mean value theorem we have,

$$\begin{aligned} L_0^\delta(P) &= \int_0^\delta W(P(t))||P'(t)||dt \\ &= \lim \Sigma\, W(P(\bar{t}_i))||P'(\bar{t}_i)||(t_{i+1}-t_i) \\ &\geq \lim \Sigma\, W(P(\bar{t}_i))(||P(t_{i+1})||-||P(t_i)||) \\ &= \int_0^\delta W(P(t))d||P(t)|| \end{aligned}$$

this last equality following from the fact that $\int_0^\delta W(P(t))d||P(t)||$ is defined since $g(t) = ||P(t)||$ is of bounded variation on $[0,\delta]$. Now we have that

$$\begin{aligned} \infty > \int_0^b W(P(t))||P'(t)||dt &\geq \int_0^\delta W(P(t))d||P(t)|| \\ &> \int_0^\delta \inf_{||x||=||P(t)||} W(x)\; d||P(t)|| \\ &= \int_0^\infty u(||P(t)||)d||P(t)|| = \int_{||P(0)||}^{||P(\delta)||} u(s)ds\,. \end{aligned}$$

Since $\int_0^\infty u(s)ds = \infty$, it follows that $\{P(t)\}_{0\leq t<b}$ is bounded. Note that $\sup\{s:u(s) > 0\} = \infty$. Since $u(s)$ is nonincreasing, we have that $W(x)$ is bounded from below on any bounded set. In particular, $W(P(t))$ is bounded from below by some number $\alpha > 0$, for all $0 \leq t < b$. Let $t_i \to b$, $(t_i \leq t_{i+1})$. Then,

$$\begin{aligned} \sum_{i=1}^n ||P(t_{i+1})-P(t_i)|| &\leq \sum_{i=1}^n \sup_{t_i \leq t \leq t_{i+1}} ||P'(t)||(t_{i+1}-t_i) \\ &\leq \int_{t_1}^{t_{n+1}} ||P'(t)||dt \\ &\leq \frac{1}{\alpha} \int_0^b W(P(t))||P'(t)||dt < \infty\,. \end{aligned}$$

Therefore we can find an x such that $P(t_i) \to x$ as $t_i \to b$. Thus $\lim P(t_i)$ exists for any increasing sequence $t_i \to b$, and is in fact unique for such sequences (since from any two such sequences we can form a new increasing sequence containing the original ones as subsequences). Hence $\lim_{t \to b} P(t)$ exists as every sequence $P(t_i)$ has a unique limit point which is independent of the sequence t_i.

Clearly F is a local homeomorphism. In view of Plastock's theorem, we need only show that F satisfies condition (L).

Suppose P(t) is defined on $0 \le t \le b$ and satisfies $F(P(t)) = L(t)$ for $0 \le t < b$. It suffices to show that F satisfies condition (L) only for those paths P(t) that are constructed in the proof of Plastock's theorem. Furthemore, if P(t) is such a path, from the local inverse function theorem, it follows that P(t) is continuously differentiable on $0 < t < b$. Since $F(P(t)) = L(t)$ on $0 \le t < b$, we use chain rule and get $J(P(t))P'(t) = L'(t) (= z \text{ say})$. That is, $P'(t) = J(P(t))^{-1}z$ for $0 \le t \le b$. Observe that,

$$\begin{aligned} L_o^b(P) &= \int_0^b W(P(t))||P'(t)||dt \\ &= \int_0^b \frac{1}{||J(P(t))^{-1}||} ||J(P(t))^{-1}z||dt \\ &\le b||z|| < \infty . \end{aligned}$$

Thus F satisfies condition (L) and consequently from Plastock's theorem, we can conclude that F is a homeomorphism. In fact it is easy to see that F^{-1} is differentiable and therefore F is a diffeomorphism of R^n onto R^n. This terminates the proof of theorem 2.

Corollary 1 : Let $F:R^n \to R^n$ be a $C^{(1)}$ differentiable map. Suppose $|\det J(x)| > \alpha > 0$ for every $x \in R^n$. Then F is a diffeomorphism of R^n onto R^n if any of the following conditions is met.

(a) $||J(x)|| \le M$ for every $x \in R^n$.

(b) F is quasi-conformal - that is there exists M such that $||J(x)|| \; ||J(x)^{-1}|| \le M$ for every $x \in R^n$.

Proof : We will prove under condition (a). A similar proof can be given under condition (b). We need the following fact for the proof : If $L:R^n \to R^n$ is an invertible linear operator then

$$|\det L| \; |(L^{-1}x,y)| \le ||x|| \; ||y|| \; ||L||^{n-1} \; (n-1)^{-(n-1)/2}$$

for all $x,y \in R^n$. [For a proof of this see Dunford and Schwartz Part II, p. 1020]. From this, we have that

$$|\det J(x)| \; |(J(x)^{-1}z,w) \le c(n)||z|| \; ||w|| \; ||J(x)||^{n-1}$$

where $c(n) = (n-1)^{-(n-1)/2}$. Let z be such that $||z|| = 1$ and $w = J(x)^{-1} z$. Then the above inequality becomes,

$$|\det J(x)| \; ||J(x)^{-1}z||^2 \leq c(n)||J(x)^{-1}z|| \; ||J(x)||^{n-1}.$$

Since $|\det J(x)| > \alpha > 0$ and $||J(x)|| \leq M$,

$$||J(x)^{-1} z|| \leq c(n)M^{n-1}/\alpha \quad \text{for all} \quad ||z|| = 1$$

That is

$$||J(x)^{-1}|| \leq c(n)M^{n-1}/\alpha .$$

It is easy to verify that the condition imposed in theorem 2 is satisfied. Thus F is a diffeomorphism of R^n onto R^n. This terminates the proof of corollary 1.

Corollary 2 [32]: A continuous local k-extension $F:R^n \to R^n$ is a homeomorphism.

Definition : Call $F:R^n \to R^n$ a local k extension if for every $z \in R^n$ there exists a neighbourhood U of z such that $||F(x)-F(y)|| \geq k||x-y||$ on U.

Proof of Corollary 2 : We will show that F satisfies condition (L). Let $(x_0,y) \in R^n \times R^n$, $L(t) = (1-t)F(x_0) + ty$, $0 \leq t \leq 1$ and $P:[0,b) \to R^n$ a path such that $F(P(t)) = L(t)$ for $0 \leq t < b$, and $P(0) = x_0$. Since F is a local k-extension, for every $0 \leq t_0 < b$ there exists $\varepsilon > 0$ such that $|t-t_0| < \varepsilon, |s-t_0| < \varepsilon$ imply $||F(P(s))-F(P(t))|| \geq k||P(s)-P(t)||$. Therefore,

$$||P(s)-P(t)|| \leq \frac{1}{k} ||F(P(s))-F(P(t))||$$

$$= \frac{1}{k} ||F(x_0)-y|| \cdot |t-s| = M \cdot |t-s| .$$

It follows that P is a local M-Lipschitz map and consequently $\lim_{t \to b} P(t)$ exists. In other words F satisfies condition (L). Also observe that F is a local homeomorphism from Brouwer's theorem on the invariance of domain. Thus F is a homeomorphism onto R^n from Plastock's theorem.

Corollary 2 is due to R.Radulescu and S.Radulescu [60]. We now have the following

Theorem 3 : Let $\{F_m\}$ be a sequence of homeomorphisms of R^n onto R^n converging uniformly to F where F is a map from $R^n \to R^n$. If F is a continuously differentiable map with non-vanishing Jacobian then F is a homeomorphism of R^n onto R^n.

Proof : Observe that each F_m is a proper map and therefore F is a proper map. Thus from Hadamard's theorem it follows that F is a homeomorphism of R^n onto R^n. However one can also prove that F is onto as follows. Let $a \in R^n$. Then for every m there exists x_m such that $F_m(x_m) = a$ for each F_m is onto R^n. Note that

$$\sup_{x \in R^n} ||F_m(x)-F(x)|| \geq ||F_m(x_m)-F(x_m)||.$$

Thus $F(x_m) \to a$ as $m \to \infty$. Since $||a|| < \infty, \{x_m\}$ is a bounded sequence. Without loss of generality let $x_m \to x_0$ as $m \to \infty$. Then $F(x_0) = a$. This yields the desired result.

We will now give examples to show the sharpness of the results obtained so far in this chapter. The first two examples indicate that theorem 2 may fail if we omit the analytic condition. Third example indicates that the full force of theorem 3 may not be valid if the convergence is not uniform.

Example 1 : Let $F(x_1,x_2) = (\tan^{-1} x_1, x_2(1+x_1^2)^2)$ be a map from R^2 to R^2. Here the Jacobian J is given by

$$J(x) = \begin{bmatrix} \dfrac{1}{1+x_1^2} & 0 \\ & \\ 4x_1x_2(1+x_1^2) & (1+x_1^2)^2 \end{bmatrix}$$

Since J is a P-matrix, F is one-one but not onto. By looking at the characteristic polynomial of J(x), it is not hard to check that $\lambda = 1/(1+x_1^2)$ is an eigenvalue, and so $\frac{1}{\lambda} = (1+x_1^2)$ is an eigenvalue of $J(x)^{-1}$. Hence $||J(x)^{-1}|| \geq 1/(1+x_1^2)$. Now

$$\int_0^\infty \inf_{||x||=t} \frac{1}{||J(x)^{-1}||} df \leq \int_0^\infty \frac{dt}{1+t^2} < \infty .$$

Thus in this example analytic condition is violated and the map F is univalent but not onto R^2.

Example 2 : Let $F(x_1,x_2) = (e^{x_1} \cos x_2, e^{x_1} \sin x_2)$. This map is neither one-one nor onto (it omits 0). Observe that $||J(x)|| = e^{x_1}$.(Here $x = (x_1,x_2)$) $1/||J(x)^{-1}|| \leq e^{x_1}$. So

$$\inf_{||x||=t} \frac{1}{||J(x)^{-1}||} \leq e^{-t} \text{ and } \int_0^\infty e^{-t} dt < \infty .$$

Example 3 : Let $f_m(x) = e^x - (e^{-x}/m)$ be a map from R^1 to R^1. Clearly each $f_m(x)$ is a homeomorphism of R^1 but the limiting function $f(x) = e^x$ is not onto R^1 though it is one-one. However convergence is not uniform over R^1.

A theorem of More and Rheinboldt and its consequences : The following theorem is due to More and Rheinboldt and the proof depends on degree theory-in particular the homotopy invariance theorem of degree theory is used. In this section as well as at several other places we use results from degree theory. A good source of reference from the point of view of mathematical analysis would be Berger [2], Cronin [13], Ortega-Rheinboldt [48], Rado-Reichelderfer [59] and Rothe [63].

Theorem 4 : Let $F:\Omega \to R^n$ be a continuously differentiable map with $J(x)$ nonsingular for every $x \in \Omega$ where Ω is an open subset of R^n. Suppose, for every $\varepsilon > 0$, the map $G_\varepsilon:\Omega \to R^n$ defined by $G_\varepsilon(x) = F(x) + \varepsilon x$ is one-one in Ω. Then F is one-one.

Proof : Suppose $F(a) = F(b) = c$ for some $a \neq b$. We can choose open sets U_a and U_b of a and b respectively with $\bar{U}_a \cap \bar{U}_b = \phi$ and $\bar{U}_a$, $\bar{U}_b \subset \Omega$. We can also choose $\bar{U}_a$ and $\bar{U}_b$ such that F restricted to $\bar{U}_a$ as well as $\bar{U}_b$ is one-one (locally one-one). In other words if $F(x) = c$ for some $x \in \bar{U}_a \cup \bar{U}_b$ it will imply $x = a$ or $x = b$. Hence $\deg(F,D,c)$ is well defined for $D = = \bar{U}_a$, $D = \bar{U}_b$ as well as $D = \bar{U}_a \cup \bar{U}_b$. (Roughly speaking $\deg(F,D,c) = \#$ of the solutions in D satisfying $F(x) = c$, $x \in D$). For any one of these three sets consider the homotopy $H_a:\Omega \times [0,1] \to R^n$ defined by $H_a(x,t) = (1-t)\,F(x) + t(x-a+c)$ (as well as the homotopy H_b defined by $H_b(x,t) = (1-t)\,F(x) + t(x-b+c)$). We will now prove that $H_a(x,t) \neq c$ for $x \in \partial D$ (= boundary of D) and $t \in [0,1]$. Plainly $H_a(x,0) = F(x) \neq c$ and $H_a(x,t) = x - a + c \neq c$ for any $x \in \partial D$. Suppose for some $t \in (0,1)$ and $x \in \partial D, H_a(x,t) = c$. This means $(1-t)F(x) + t(x-a+c) = c$. This implies that $F(x) + (\frac{t}{1-t})x = c + (\frac{t}{1-t})a = F(a) + (\frac{t}{1-t})a$ but this contradicts the assumption that $F(x) + \varepsilon x$ is one-one. Hence $H_a(x,t) \neq c$, $H_b(x,t) \neq c$ for $x \in \partial D$, $t \in [0,1]$. Invoking the homotopy invariance theorem of degree theory, (see chapter VI in [48]) it follows that

$$\begin{aligned} \deg(F,D,c) &= \deg(x-a+c,D,c) \\ &= \deg(x-b+c,D,c) . \end{aligned}$$

This immediately tells us that $\deg(F,\bar{U}_a,c) = \deg(F,\bar{U}_b,c) = \deg(F,\bar{U}_a \cup \bar{U}_b,c) = 1$. But by assumption $\deg(F,\bar{U}_a \cup \bar{U}_b,c) = \deg(F,\bar{U}_a,c) + \deg(F,\bar{U}_b,c)$ since $\bar{U}_a \cap \bar{U}_b = \phi$ and this leads to a contradiction. Hence F is univalent on Ω and this terminates the proof of the theorem.

Another proof of theorem 4 (due to Mas-Colell) : Suppose F is not as desired, that is, $F(a) = F(b) = c$, $a \neq b$. Take U_a, U_b with $U_a \cap U_b = \phi$, such that for small ε, there are functions $x_a(\varepsilon)$, $x_b(\varepsilon)$ satisfying $F(x_a(\varepsilon))) + \varepsilon x_a(\varepsilon) = c$ and $F(x_b(\varepsilon)+\varepsilon x_b(\varepsilon)=c$. These functions exist locally by the implicit function theorem and the regularity of $J(a)$, $J(b)$ and we have a contradiction for $x_a(\varepsilon) \neq x_b(\varepsilon)$. See also pp 218-222 in [2]. The following corollary follows from the fundamental theorem of Gale-Nikaido and

Theorem 4.

Corollary 3 : Let $\{F_m\}$ be a sequence of differentiable functions defined over Ω where Ω is an open rectangle in R^n, and let the Jacobian matrix for each n and x, be a P-matrix. Suppose $F_m(x)$ converges to $F(x)$ for each $x \in \Omega$. Further suppose F is differentiable with nonvanishing Jacobian for each $x \in \Omega$. Then F is one-one on Ω.

Remark : Let $\Omega = R^n$ and suppose each F_m is onto R^n. Is F also onto? The following simple example shows that it need not be. Let $\Omega = R^1$ and $F_m(x) = \tan^{-1} x + x/m$. Clearly each F_m is a homeomorphism onto R^1 but the limiting function $F(x) = \tan^{-1} x$ is one-one but not onto R^1.

Proof : From Gale-Nikaido's fundamental theorem it follows that each F_n is a P-function. Since F_n converges to F pointwise, $F + \varepsilon I$ is a P-function for each $\varepsilon > 0$. In other words $F + \varepsilon 1$ is injective for each $\varepsilon > 0$ and furthermore for each $x \in \Omega$ there is an open set U_x such that $F(y) = F(x)$ for each $y \in U_x$ implies that $u = x$. Hence we can conclude from the proof of theorem 4 that F is one-one on Ω.

We need the following definition for the next theorem.

Definition : Call a matrix A weakly positive quasidefinite if $\det A > 0$ and the quadratic from $\langle Ax,x\rangle$ is positive semidefinite. A weakly negative quasidefinite matrix is defined analogously.

Theorem 5 : Let $F:\Omega \to R^n$ be a continuously differentiable map with Ω an open convex region in R^n. Further suppose Jacobian matrix $J(x)$ is weakly positive quasidefinite everywhere. Then F is one-one.

Proof : Let $G_\varepsilon(x) = F(x) + \varepsilon x$ where $\varepsilon > 0$, it is easy to see that J_ε is positive quasidefinite. Now theorem 5 is a consequence of Gale-Nikaido's result and theorem 4 of the present chapter.

Light open mappings and homeomorphisms : In this section we will prove two results due to McAuley [38] where we place purely topological conditions under which the map F becomes a homeomorphism. We do not even assume F to be differentiable. While Gale-Nikaido's result depends on the choice of a fixed coordinate system, the problem under consideration namely "$F|\Omega$ is univalent" does not depend on the choice of a fixed coordinate system. McAuley's result eliminates this particular difficulty.

We will start with a few definitions. Throughout this section we will assume F to be a continuous map from $\Omega \to R^n$.

Definition : Call F a light mapping if $F^{-1}(F(x))$ is totally disconnected for each $x \in \Omega$.

Definition : Call F an open map if for each U open in Ω, F(U) is open relative to $F(\Omega)$.

Remark 1 : Local homeomorphisms are special light mappings while covering mappings are indeed open local homeomorphisms. The latter are simple cases of light open mappings.

Remark 2 : The study of light open mappings has deep motivations in the topological nature of analytic functions defined in domains in the complex plane. For example a non-constant function $w = g(z)$, analytic in a region D of the (complex) z-plane which takes D into the w-plane is strongly open. That is if U is open in D, then g(U) is open in the w-plane. Furthermore, g is light. In fact the set $g^{-1}g(x)$ is discrete (has no limit point). For more details see [38].

Let F be a light mapping from Ω to R^n. We shall say that the singular set S_F of F is the set of points $x \in \Omega$ such that F is not locally one-one at x, that is, there is no set U open in Ω and containing x such that $F|U$ is one-one. Some authors [10] refer to S_F as the branch set of F. The singular set S_F is closed. We shall now give two sufficient conditions under which a light mapping F of Ω onto $F(\Omega)$ is open.

(i) Floyd's condition : For each continuous mapping $g:[0,1] \to F(\Omega)$ and for each $x \in F^{-1}(g(o))$, there is a continuous map $h:[0,1] \to \Omega$ such that $g = F \circ h$ and $h(0) = x$.

(ii) McAuley's condition : The singular set S_F of F has the property that $F(S_F)$ fails to separate any set V open relative to $F(\Omega)$ and $F(S_F)$ contains no nonempty open set.

For a proof of (i) see Floyd [Ann. Math. Vol. 51,(1950)] and for a proof of (ii) see [38].

Our aim in this section is to provide conditions under which a light open mapping F of a compact subset Ω of R^n into R^n is a homeomorphism. It is not difficult to check that F(int Ω) = int Ω and F(boundary Ω) = boundary $F(\Omega)$ if such a mapping were to be a homeomorphism. We are ready to state our theorems.

Theorem 6 : (McAuley). Suppose Ω is a compact set in R^n with $\partial\Omega \neq \phi$ and int $\Omega \neq \phi$ and F is a continuous light open mapping of Ω into R^n such that (1) F(int Ω) = int $F(\Omega)$, (2) $F(\partial\Omega) = \partial F(\Omega)$, (3) the singular set S_F has the property that $F(S_F)$ does not contain a non-empty set open relative to $F(\Omega)$, (4) $F(S_F)$ does not separate $F(\Omega)$, and (5) there exists a non-empty U in Ω open relative to Ω such that

$F|U$ is one-one and $F^{-1}F(U) = U$. Then F is a homeomorphism.

Theorem 7 (McAuley) : Suppose that F is a continuous light open mapping from a unit ball Ω in R^n onto another unit ball B in R^n such that (1) $F^{-1}F(\partial\Omega) = \partial\Omega$, (2) $F(\partial\Omega) = \partial B$, (3) $F|S_F$ is one-one, and (4) for each component C of $B \setminus F(S_F)$ there is a nonempty V in C open relative to B such that $F|F^{-1}(V)$ is one-one. Then F is a homeomorphism.

Remark 1 : If F is a differentiable map with non-vanishing Jacobian, then $S_F = \phi$. Furthermore if Ω is compact then clearly $F(\text{int } \Omega) = \text{int } F(\Omega)$ and $F(\partial\Omega) = \partial F(\Omega)$ - for a proof of this assertion see (p 447-448 in [38]).

Remark 2 : It is not clear whether Gale-Nikaido's univalent theorem proved in the last chapter is a consequence of theorem 6. In fact if F and Ω satisfy the conditions imposed on Gale-Nikaido's theorem, it is easy to verify that conditions (1),(2),(3) and (4) given in theorem 6 are met. However it is not at all obvious how to check condition (5) of theorem 6.

Proof of theorem 6 : Let P denote the set of all y in $F(\Omega)$ such that $F^{-1}(y)$ is nondegenerate. It is clear that F is a homeomorphism iff P is empty. Suppose P is not empty. It is not hard to check that the set P is open relative to $F(\Omega)$ and contains a non-empty open set since F is an open map.

Let $A = F(S_F) \cup P$. We will prove that A is closed and consequently compact. Suppose that y is a limit point of A but $y \in F(\Omega) \setminus A$. Since $F(S_F)$ is closed, it follows that y is a limit point of $P \setminus F(S_F)$. Clearly $F^{-1}(y)$ is degenerate and let $x = F^{-1}(y)$. Also F is locally one-one at x. In other words there exists a neighbourhood N_x of x such that $F|N_x$ is one-one, $F(N_x)$ is open relative to $F(\Omega)$, and $F^{-1}F(N_x) = N_x$. Thus $F(N_x) \cap (P \setminus F(S_F)) = \phi$. This contradicts the fact that y is a limit point of $P \setminus F(S_F)$. Thus the set A is closed and consequently compact.

Since $F(S_F) \cup P$ is closed, we have $B = F(\Omega) \setminus (F(S_F) \cup P) = (F(\Omega) \setminus F(S_F)) \cap (F(\Omega) \setminus P)$ is open in $F(\Omega)$. From condition (5) of theorem 6, it follows that each of B and $F(\Omega) \setminus F(S_F)$ is nonempty. Observe that from condition (4), $F(\Omega) \setminus F(S_F)$ is connected. If we show that $F(\Omega) \setminus F(S_F) \cap P$ is open relative to $F(\Omega)$ then we arrive at a contradiction to the fact that $F(\Omega) \setminus F(S_F)$ is connected. Hence it will follow that $(F(\Omega) \setminus F(S_F)) \cap P = \phi$ and condition (3) will now imply that $P = \phi$, contrary to our assumption that P is not empty. Thus the proof of the theorem 6 will be complete if we show that $(F(\Omega) \setminus F(S_F)) \cap P$ is open relative to $F(\Omega)$.

Since $F(\text{int } \Omega) = \text{int } F(\Omega)$ and $F(\partial\Omega) = \partial F(\Omega)$, we have that $F(S_F) \cap \text{int } F(\Omega) = F(S_F \cap \text{int } \Omega)$. Now, $F(S_F) = F(S_F \cap \partial\Omega) \cup F(S_F \cap \text{int } \Omega)$. Since P is open in $F(\Omega)$ and $F(S_F)$ is closed, it follows that $(F(\Omega) \setminus F(S_F)) \cap P$ is open relative to $F(\Omega)$. This terminates the proof of theorem 6.

The following corollary which is needed for the proof of theorem 7 is an immediate consequence of theorem 6.

Corollary 4 : Suppose Ω is a compact subset of R^n with int $\Omega \neq \phi$, Ω = closure of (int Ω), int $F(\Omega)$ is connected, and that F is a light open mapping of Ω into R^n such that (1) $F|(\text{int } \Omega)$ is locally one-one, (2) $F(\text{int } \Omega) = \text{int } F(\Omega)$, (3) $\partial F(\Omega)=F(\partial\Omega)$, and (4) there is U open relative to Ω such that $F^{-1} F(U) = U$ and $F|U$ is one-one. Then F is a homeomorphism.

Proof : It is not hard to check that $\partial\Omega \supset S_F$, $F(\Omega) \setminus F(S_F)$ is connected, and $F(S_F)$ contains no nonempty set open relative to $F(\Omega)$. The hypothesis of theorem 6 is satisfied and the corollary follows.

Proof of theorem 7 : Conditions (1) and (2) imply that, $F(\text{int } \Omega) = \text{int } B$. Since B is locally connected, we can find at most a countable number of components $C_1, C_2, C_3, \ldots$, of $B-F(S_F)$ and moreover, if there is an infinite number of components, then diameter $C_i \to 0$, that is, the sequence $\{C_i\}$ is a null sequence. For each i, $F^{-1}(C_i)$ is connected. Let $K_i = F^{-1}(C_i)$. Note that $F(K_i) = C_i$. These facts follow from condition (4) of theorem 7 and a theorem of Whyburn ([72], (7.5); p. 148). Moreover $\{K_i\}$ is a null sequence.

Observe $F|K_i$ is locally one-one since $K_i \cap S_F = \phi$. Furthermore, $F(\bar{K}_i) = \bar{C}_i$ (where $\bar{D}$ denotes the closure of D), and $F^{-1}(\bar{C}_i) = \bar{K}_i$. Also it is not hard to check that, $F(\partial\bar{K}_i) = \partial F(\bar{K}_i), \partial F(\text{int } \bar{K}_i) = \text{int } F(\bar{K}_i)$, and $F|\bar{K}_i$ is a light open mapping of $\bar{K}_i$ onto $\bar{C}_i$.

Apply corollary 4 to each of $\bar{K}_i$. Thus, $F|\bar{K}_i$ is a homeomorphism. Set $S=\cup\bar{K}_i$. Since $F|S_F$ is one-one, S_F contains no set open relative to Ω (the unit ball in R^n in the hypothesis), indeed, $F(S_F)$ contains no set open relative to B. Note that each point of $F(S_F)$ is a limit point of $B-F(S_F)$. Thus $B = \cup \bar{C}_i$. It follows that $F(S) = B$ and $F|S$ is a homeomorphism of S onto B. We conclude that $S = \Omega$ and that F is a homeomorphism. This terminates the proof of theorem 7.

Remark 1 : Condition (5) in theorem 6 (as well as condition (4) in theorem 7) is crucial for the proof of theorem 6 (theorem 7). Also it appears difficult to check this condition in a given problem.

Remark 2 : In order to apply these results it is not even necessary unlike Gale-Nikaido's theorem for F to be differentiable (everywhere). These results hold good in more general spaces - see [38]. The following example shows that condition (5) is crucial for the proof of theorem (6). Without this condition theorem 6 may fail.

Example : Let $F = (f,g)$ with $f(x,y) = x^2-y^2$ and $g(x,y) = 2xy$. Let $\Omega = \{(x,y) : x^2+y^2 \leq 1\}$. Here the Jacobian of F is given by

$$J = \begin{bmatrix} 2x & -2y \\ 2y & 2x \end{bmatrix}$$

and $\det J = 4(x^2+y^2) \neq 0$ iff $(x,y) \neq (0,0)$. Clearly origin is the only point where F is not locally one-one. In other words $S_F = \{(0,0)\}$. Also it is not hard to check conditions (1) - (4) of theorem 6 are satisfied, but condition (5) is not satisfied.

Remark 3 : One can prove the following result using theorem 6. Suppose Ω is a compact subset of R^n with int $\Omega \neq \phi$. Furthermore, suppose F is a local homeomorphism of Ω into R^n such that (i) $F(\Omega)$ is connected and (ii) there is some set U in Ω open relative to Ω such that $F^{-1}F(U) = U$ and $F|U$ is one-one. Then F is a homeomorphism. It is clear from this remark as well as corollary 4 that one can derive extremely useful results from theorem 6 (as well as theorem 7). We will again make use of this result in the last chapter.

We will close this chapter with an old conjecture of Whyburn. Suppose F is a continuous light open mapping of $I^n = [0,1] \times [0,1] \times \ldots \times [0,1]$ (cartesian product of unit interval taken n times) onto I^n such that $F^{-1} F(\partial I^n) = \partial I^n$ and $F|\partial I^n$ is one-one. Does it follow that F is a homeomorphism? The answer is 'yes' when $n = 2$. But the answer is not known for $n \geq 3$. For more details regarding this conjecture see McAuley [38] or Whyburn [74]. However this conjecture is true (via Kestelman's result quoted in the previous chapter) if we further suppose that (i) F is continuously differentiable and (ii) the Jacobian matrix of F is non-singular for every $x \in I^n$.

CHAPTER V

SCARF'S CONJECTURE AND ITS VALIDITY

Abstract : In this chapter we will prove a substantial generalization of Gale-Nikaido's theorem on univalent mappings in which we assume P-property only on the boundary of the rectangular region and this was conjectured by Scarf. We will give two different proofs one due to Garcia and Zangwill and the other due to Mas-Colell. Proof of Garcia and Zangwill uses the norm-coerciveness theorem whereas Mas-Colell uses results from degree theory. There is a subtle difference between these results and Gale-Nikaido's fundamental theorem. The difference lies in the fact that the proofs of Garcia-Zangwill and Mas-Colell demand F to be a $C^{(1)}$ function whereas Gale-Nikaido's result holds good if we assume F to be a differentiable function not necessarily a $C^{(1)}$ function. It is not clear whether Garcia-Zangwill or Mas-Colell's result holds good if we assume F to be a differentiable function. This seems to be an interesting open problem in this area. Another problem which remains still unanswered is the following: Suppose F is a $C^{(1)}$ map from a compact convex set $\Omega \subset R^n$ to R^n. Suppose Jacobian of F is a P-matrix for every $x \in \Omega$. Does it imply F is one-one in Ω ?

Garcia-Zangwill's result on univalent mappings : In order to state the result we need the following definition.

Definition : Let Ω be a compact rectangle in R^n with boundary $\partial\Omega$ and F a map from Ω to R^n. Call F norm-coercive if $||F(x)|| \to \infty$ as x approaches $\partial\Omega$. The following theorem is known and a proof may be found in [48, p. 136]. Let Ω^o stand for the interior of Ω.

Norm-Coerciveness Theorem : Let $F:\Omega^o \to R^n$ be a continuously differentiable norm-coercive mapping with det J(x) positive for every $x \in \Omega$. Then F is one-one (and in fact F is a homeomorphism).

Remark : In finite dimensional spaces, concept of norm-coerciveness in equivalent to the fact that the map is proper. For a proof see Berger [2].

We will now state and prove (Scarf's Conjecture).

Theorem 1 : Let $F:\Omega \to R^n$ be continuously differentiable on a bounded rectangle Ω with det $J(x) > 0$ for every $x \in \Omega$. Further suppose J(x) is a P-matrix for every $x \in \partial\Omega$ (= boundary of Ω). Then F is univalent on Ω. [We will assume $\Omega = \{x:a_i \leq x_i \leq b_i$

for $i = 1,2,\ldots,n\}]$.

Idea of the proof can be explained simply as follows. We will construct a function G which will coincide with F except near the boundary $\partial\Omega$. Also this function G will be norm coercive. We will use norm-coerciveness theorem to conclude that G is univalent which in turn will imply that F is also univalent. One such G can be defined as follows:

$$g_i(x) = f_i(x)+d_i(x_i) \quad \text{for} \quad i = 1,2,\ldots,n$$

where

$$d_i(x_i) = -\left(\max(0, \frac{1}{x_i-a_i} - \frac{1}{\delta_1})\right)^2 + \left(\max(0, \frac{1}{b_i-x_i} - \frac{1}{\delta_1})\right)^2 .$$

Here δ_1 is any positive number less than $\min\left(\frac{b_i-a_i}{2}\right)$. It is not hard to see that the function $d_i(x_i)$ has the following properties:

(i) $d_i(x_i) = 0$ if $a_i + \delta_1 \le x_i \le b_i - \delta_1$

(ii) $d_i'(x_i) > 0$ if $a_i < x_i < a_i + \delta_1$ or $b_i - \delta_1 < x_i < b_i$ and

(iii) $|d_i(x_i)| \to \infty$ if either $x_i \downarrow a_i$ or $x_i \uparrow b_i$.

Also it is easy to see that $G = (g_1,g_2,\ldots,g_n)$ coincides with F for all x which are at least a distance δ_1 from the boundary. Also G is norm coercive. Further since f_i's are continuously differentiable functions and Ω is compact we can find a δ_2 such that $J(x)$ will be a P-matrix for all x which are at most a distance δ_2 from the boundary. Define $\delta = \min\{\delta_1,\delta_2\}$. Now we prove theorem 1.

<u>Proof of theorem 1</u> : Suppose $x \neq y$ with $F(x) = F(y)$ then we will arrive at a contradiction. We will analyse two cases (i) x and y are interior points, (ii) at least one of them is a boundary point. In fact case (ii) can be reduced to case (i) by modifying F slightly as will be shown below. We will now prove that case (i) is impossible. Construct a G as described above. Since δ_1 as well as δ_2 can be chosen as small as we please, we will choose them so small so that both x and y are more than a distance of δ from the boundary. This will imply $G(x) = G(y)$ since $G(x) = F(x)$, $G(y) = F(y)$ and $F(x) = F(y)$. Clearly G is continuously differentiable and it is norm coercive. If $x \in \Omega$ is at least a distance of δ from $\partial\Omega$, Jacobian of $G(x)$ = Jacobian of F and hence det of the Jacobian of $G(x)$ is positive. If x is within δ of $\partial\Omega$, Jacobian $G(x)$ is a P-matrix and hence its determinant is positive. In other words conditions of the norm coerciveness theorem are satisfied and hence G should be one-one which contradicts $G(x) = G(y)$ for $x \neq y$. Hence F must be one-one. We will now demonstrate that case (ii) can be reduced to case (i). Suppose $F(x) = F(y)$ for two distinct points x and y in Ω not both in Ω^o. Choose x' and y' in Ω^o close to x and y respectively. Define $H:\Omega \to R^n$ as follows and its ith component is given by

$$h_i(x) = f_i(x) - \frac{b_i(x_1 - y_1')}{(x_1' - y_1')} - \frac{b_i'(x_1 - x_1')}{(y_1' - x_1')}$$

where $F(x') = b$ and $F(y') = b'$. Clearly H is continuously differentiable and

$$\frac{\partial h_i(x)}{\partial x_j} = \begin{cases} \dfrac{\partial f_i(x)}{\partial x_j} & \text{if } j \neq 1 \\[2ex] \dfrac{\partial f_i(x)}{\partial x_1} + \dfrac{b_i - b_i'}{y_1' - x_1'} & \text{if } j = 1 . \end{cases}$$

Since we can make the term $(b_i - b_i')/(y_1' - x_1')$ arbitrarily close to zero by selecting x' and y' close to x and y respectively it follows that H will satisfy the conditions of theorem 1. Also observe that $H(x') = H(y')$ which is impossible from case (i). This terminates the proof of theorem 1.

Remark : During the course of the proof of theorem 1 we have made use of the following result: If D_I is an n x n diagonal matrix with non-zero diagonal entries only in positions $i \in I$ and if P is an arbitrary n x n matrix then determinant $(P + D_I) = \sum_{K \subset I} (\prod_{k \in K} d_k) M_K$ where M_K's are the minors of the matrix P. Here M_K is the minor resulting from P omitting the rows i and columns i for $i \in K$. Also $M_\phi = \det P$ and $M_{\{1,\ldots,n\}} = 1$ by definition. As this formula may be a bit confusing. Here is an example for $n = 3$, $I = \{1,2,3\}$.

$$\det(P + D_I) = \det P + d_1 M_1 + d_2 M_2 + d_3 M_3 + d_1 d_2 M_{12} + d_2 d_3 M_{23} + d_1 d_3 M_{13} + d_1 d_2 d_3$$

We make use of this result to conclude that the determinant of the Jacobian matrix of G is positive near the boundary. This is where the proof will breakdown if we imitate this to show that a map F will be one-one if we assume that the Jacobian is an N-matrix for all $x \in \partial\Omega$ in theorem 1. This problem appears to be an interesting open problem. We believe that theorem 1 should be true under the assumption that the Jacobian is an N-matrix for all $x \in \partial\Omega$ and that the determinant of the Jacobian is everywhere negative.

Note that theorem 1 is a generalization in a certain direction of Gale Nikaido's fundamental theorem when F is continuously differentiable function with its Jacobian everywhere a P-matrix. However one should also note that in Gale Nikaido's theorem we do not assume that the partial derivatives to be continuous as in Theorem 1. It is not clear whether theorem 1 remains true when F is a differentiable but not a $C^{(1)}$ function.

From theorem 1 one can deduce the following:

Theorem 2 : Let $F : R^n \to R^n$ be continuously differentiable. Let $\Omega \subset R^n$ be a bounded rectangle. Suppose that the determinant $J(x) > 0 \; \forall \; x \in R^n$ and further suppose $J(x)$ is a P-matrix $\forall \; x \in R^n \setminus \Omega$. Then F is one one.

Proof : Suppose $F(x) = F(y)$ with $x \neq y$. Let Ω' be a compact rectangle containing x,y and Ω in its interior. Then F satisfies all the conditions of theorem 1 on Ω' and consequently F is one one on Ω' contradicting our original assumption. This terminates the proof of theorem 2.

At this juncture it is worth mentioning that Garcia and Zangwill prove theorem 1 (as well as theorem 2) under slightly weaker assumptions on the Jacobian matrix on the boundary. Let $I(x) = \{i : x_i = a_i \text{ or } x_i = b_i\}$. We say that F has the S property if $J(x)_K \, (\prod_{i \in K} J(x)_i) > 0$ for all $x \in \Omega$ and $K \subset I(x)$. Garcia and Zangwill prove univalence when F satisfies the S property. This S property permits certain principal minors to be negative on the boundary. We will now give an example of a one-one function F whose Jacobian is not everywhere a P-matrix. Let $F:R^2 \to R^2$ be a function defined as follows: $F = (f_1, f_2)$, where $f_1(x,y) = x^3/3 + x(y^2 - \frac{1}{2}) - y$ and $f_2(x,y) = x+y$. Then

$$J(x,y) = \begin{bmatrix} x^2 + y^2 - \frac{1}{2} & 2xy - 1 \\ 1 & 1 \end{bmatrix}$$ Plainly $J(x,y)$ is not a P-matrix.

Also $\det J(x,y) = (x-y)^2 + \frac{1}{2} > 0 \; \forall \; (x,y) \in R^2$. However $J(x,y)$ is a P-matrix if $x^2+y^2 > \frac{1}{2}$. Consequently F satisfies the conditions of theorem 2 and hence F is one-one throughout R^2.

Mas-Colell's univalent result : We will now prove the following theorem of Mas-Colell.

Theorem 3 : Let $F:\Omega \to R^n$ be a continuously differentiable function where Ω is a compact convex polyhedron of full dimension. For every nonempty subspace $L \subset R^n$ let $\Pi_L : R^n \to L$ denote the perpendicular projection map. If for every $x \in \Omega$ and subspace $L \subset R^n$ spanned by a face of K (that is, the translation to the origin of the minimal affine space containing K)which includes x, the map $\Pi_L \, DF(x) : L \to L$ has a positive determinant (that is the linear map $\Pi_L \cdot DF(x)$ preserves orientation where $DF(x)$ stands for the derivative map of F at x) then F is one-one on Ω and consequently a homeomorphism.

Several remarks will be in order now. Here Ω is assumed to be simply a compact convex polyhedral set. As such this theorem includes Gale-Nikaido's fundamental theorem on univalence mappings. Conditions imposed on this theorem are coordinate free, in the sense that their formulation does not rely on a previous choosing of coordinates. Proof of this theorem will depend on the following known results from degree theory.

Theorem (a) : Let $F, G: \bar{C} \subset R^n \to R^n$ be two continuous maps where $\bar{C}$ is the closure of an open bounded set C. Suppose there exists a homotopy $H: \partial C \times [0,1]$ such that $H(x,0) = F(x)$, $H(x,1) = G(x)$ for all $x \in \partial C$. If $y \in R^n$ such that $H(x,t) \neq y$ for all $x \in \partial C$, $t \in [0,1]$, then $\deg(F,C,y) = \deg(G,C,y)$.

For a proof see Schwartz 1964, p. 93 Non-linear functional analysis, Lecture notes Courant Inst. of Math. Sci., New York.

Theorem (b) : Let $F: D \subset R^n \to R^n$ be continuously differentiable on the open set D and C an open bounded set such that $\bar{C} \subset D$. Given a coordinates system define $A = \{x \in \bar{C}$, Jacobian at x is singular$\}$. If $y \notin F(\partial C \cup A)$ then either $\Gamma = \{x \varepsilon c | F(x) = y\}$ is empty and $\deg(F,C,y) = 0$ or Γ consists of finitely many points $x^1, \ldots, x^m$ and

$$\deg(F,C,y) = \sum_{j=1}^{m} \text{sign} \det J(x^j).$$ (For a proof see p. 159, [48]).

As in theorem 1, it suffices to prove that F is one-one restricted to Ω^o = interior of Ω; otherwise we can always enlarge the domain of F which will contain Ω in its interior and similar to Ω and satisfying all the other conditions of theorem 3. For every $x \in R^n$ let $s(x) \in \Omega$ be the foot of x, that is, s(x) is the unique element of minimizing $||x-s||$ for $s \in \Omega$. Plainly $s(x) = x$ for $x \in \Omega$. We now extend $F: \Omega \to R^n$ to the whole of R^n by letting a function $\tilde{F}: R^n \to R^n$ be defined as $\tilde{F}(x) = F(s(x)) + x - s(x)$. For any $y \in F(\Omega)$ define $\tilde{F}_y(x) = \tilde{F}(x) - y$. We will now state and prove two lemmas that are needed for a proof of theorem 3.

Lemma 1 : Let $S_r = \{x: ||x|| = r\}$ and $B_r = \{x: ||x|| < r\}$ be the sphere and ball of radius r. Then for any $y \in F(\Omega)$, and r sufficiently large, $\tilde{F}_y$ restricted to S_r has degree one, that is, it is homotopic to the identity in S_r with respect to R^N $R^n \setminus \{0\}$.

Proof : Clearly the homotopy bridge H(x,t) is given by $H(x,t) = t\,\tilde{F}_y(x) + (1-t)I(x)$ where $t \in [0,1]$, $I(x) = x$. We have to only check that $H(x,t) \neq 0$ for any $t \in [0,1]$ and $x \in S_r$, if r is sufficiently large. We will verify this by simply showing that $x.\tilde{F}_y(x) > 0$ for any $y \in F(\Omega)$ and $x \in S_r$ when r is sufficiently large. In fact choose any $r > \max_{z \varepsilon \Omega, y \varepsilon F(\Omega)} ||F(z)-z-y|| = s$. Then

$$x.\tilde{F}_y(x) = ||x||^2 - x.\,(s(x) + y - F(s(x)))$$
$$\geq ||x||^2 - ||x||\ \ ||s(x) + y - F(s(x))|| \geq r^2 - rs > 0.$$

This finishes the proof of lemma 1.

Lemma 2 : Let K^* be a polyhedron and F satisfy the hypothesis of theorem 3. Let $A = \{x \in R^n : \tilde{F}$ is not continuously differentiable at $x\}$. Then if $x \notin A$, $|D\tilde{F}(x)|$ (= determinant of the linear map $D\tilde{F}(x)$) is positive.

Proof : Let $x \notin A$. Then $x-s(x)$ is perpendicular to a single face of Ω, which, of course, includes $s(x)$. Let L be the subspace spanned by this face and $L^{\perp}$ the subspace orthogonal to L. For small $v \in L$, $s(x+v) = s(x)+v$ and so $\tilde{F}(x+v) = F(s(x)+v)+x+v-s(x)$; hence $D\tilde{F}(x)v = DF(s(x))v$. For $v \in L^{\perp}$, $s(x+v) = s(x)$ and so, $\tilde{F}(x+v)=F(s(x))+x+v-s(x)$. Consequently $D\tilde{F}(x)v = v$. Choose an orthogonal coordinate system whose k first coordinates generate L, $\tilde{J}(x)$, the matrix of $D\tilde{F}$ with respect to this coordinate system, takes the form

$$\tilde{J}(x) = \begin{bmatrix} J_k\, F(s(x)) & \begin{matrix} 0 \\ I \end{matrix} \end{bmatrix}$$

where $J_kF(s(x))$ are the first k columns of $J(x)$ (= Jacobian matrix for F). Therefore $|D\tilde{F}(x)| = |\tilde{J}(x)| = |J_{kk}\, F(s(x))|$ where $J_{kk}\, F(s(x))$ are the first k rows of $J_kF(s(x))$. However $J_{kk}F(s(x))$ is the matrix of $\Pi_L \cdot DF(s(x)) : L \to L$ and by hypothesis $|J_{kk}F(s(x))|>0$ and this terminates the proof of the lemma.

We are now set to prove theorem 3.

Proof of theorem 3 : Let A be the set (as defined in lemma 2) of those points $x \in R^n$ at which $\tilde{F}$ is not continuously differentiable. This set contains no open set. This is a consequence of the fact that $s(x)$ is continuously differentiable except at those x which are contained in a finite number of hyperplanes. Since F is lipschitzian, $\tilde{F}(A)$ contains no open set.

Choose $r > 0$, sufficiently large so that lemma 1 holds good. Also suppose $\tilde{F}_y^{-1}(0) \cap A = \phi$. In other words we are assuming $\tilde{F}_y^{-1}(0)$ lies entirely in the region of differentiability. From theorem (a) and theorem (b) it follows that

$$\begin{aligned} \deg(\tilde{F}_y, B_r, 0) &= \deg(I, B_r, 0) \\ &= \Sigma \text{ sign det } J\,\tilde{F}_y(x) = 1. \end{aligned}$$

Here the summation is extended over those x which belong to B_r with $\tilde{F}_y(x) = 0$. Note that $\tilde{F}_y^{-1}(0) \cap B_r \neq \phi$ for any $y \in F(\Omega)$. Also note that we are writing det $J\tilde{F}_y(x)$. [As such we will assume for this proof a coordinate system is given to us]. Invoking lemma 2, we infer $\tilde{F}_y^{-1}(0) \cap B_r$ is a singleton set.

As remarked earlier we will only show that F restricted to interior $\Omega = \Omega^o$ is one-one. Suppose there are $x_1, x_2 \in \Omega^o$ with $x_1 \neq x_2$ and $F(x_1) = F(x_2)$. One can find disjoint open sets $U_1 \ni x_1$, $U_2 \ni x_2$ with $F(U_1) \cap F(U_2) \neq \phi$ open. Since $\tilde{F}(A)$ contains no open set there is a $y \in F(U_1) \cap F(U_2)$ such that $y \notin \tilde{F}(A)$. Consequently

$\tilde{F}_y^{-1}(0) \cap A = \phi$. Note that $F^{-1}(y) \subset \tilde{F}_y^{-1}(0) \cap B_r$ and the latter set is not a singleton set as $F^{-1}(y)$ has at least two elements. This contradiction establishes that F restricted to Ω^o must be one-one and this terminates the proof of theorem 3.

Mas-Colell's proof of theorem 3 makes use of Poincare-Hopf index theorem [see pages 35-41 John Milnor's Topology from the differentiable view point (1965), the University Press Virginia]. Instead we use theorem (a) and theorem (b). We now have the following theorem due to Mas-Colell.

Theorem 4 : Let $\Omega \subset R^n$ be a compact, convex set of full dimension with smooth boundary $\partial\Omega$ (= C^1 boundary) and $F : \Omega \to R^n$ a continuously differentiable function. For each $x \in \partial\Omega$, let T_x denote the tangent plane at x. If the Jacobian $J(x)$ has a positive determinant at each $x \in \Omega$ and if for each $x \in \partial\Omega$, $J(x)$ is positive quasi-definite on T_x (that is $\langle v, J(x)v \rangle$ is positive for every $v \in T_x$ with $v \neq 0$), then F is one one (and consequently a homeomorphism).

Proof : We will first show that if $x \in \partial\Omega$ and $L \subset T_x$ is a subspace then $\Pi_L \cdot DF(x):L \to L$ has a positive determinant. In other words this theorem is a consequence of theorem 3. Assume that we are given an orthogonal coordinate system such that the first k coordinates generate L and the n-th is perpendicular to T_x and let $J(x)$ denote the Jacobian matrix. Then $J_{n-1,n-1}(x)$, the matrix obtained by omitting the n-th row and n-th column (by hypothesis) is positive quasidefinite. However, any such matrix is a P-matrix [see examples of P-matrices in chapter II]. This applies to $J_{kk}(x)$ the matrix of $\Pi_L \cdot DF(x) : L \to L$ and yields the fact that $J_{kk}(x)$ has a positive determinant. Now one can complete the proof by approximating Ω by a polyhedran Ω' and the above fact implies that the hypotheses of theorem 3 are satisfied for Ω'. Hence F is 1-1 on Ω. This completes the proof of theorem 4.

Remark 1 : Will theorem 4 be true if one assumes that $J(x)$ is weakly positive quasi-definite on T_x ? It is easy to see that the map F will be one-one on Ω^o = interior of Ω, by noting that the maps $G_\varepsilon = F(x) + \varepsilon I(x)$ are one-one over Ω where $\varepsilon > 0$ and $I(x) = x$. Also the map G_ε will be one-one throughout Ω including the boundary. It is not clear whether F is one-one throughout Ω. In other words this raises the following question: Let Ω be a closed convex region with nonempty interior and let $F:\Omega \to R^n$ be a one-one map throughout Ω^o = (interior of Ω). What other conditions should be imposed on $\partial\Omega$ and F so that F will be one-one throughout Ω. [see Kestelman's result given in chapter III]. In fact what we want is an approximation theorem similar to theorem 4 in chapter IV which will hold good for closed region Ω (with nonempty interior). In particular we would like to know whether theorem 4 holds good when $J(x)$ is weakly positive quasidefinite on T_x.

Remark 2 : Mas-Colell's conditions on $J(x)$ at the boundary of Ω neither imply nor are implied by the fact $J(x)$ is a P-matrix.

One can give a slight generalization of theorem 2. For that we need the following

Definition : Let A be an $n \times n$ matrix. Call A a weak P-matrix if $\det A > 0$ and every other principal minor is non-negative.

Theorem 2' [53] : Let $F:R^n \to R^n$ be continuously differentiable. Let Ω be a bounded rectangle in R^n. Suppose that the determinant $J(x) > 0$ for every $x \in R^n$ and further suppose $J(x)$ is a weak P-matrix for all $x \in R^n \setminus \Omega$. Then F is one-one.

Proof : We will assume without loss of generality that Ω is a compact rectangle. Let $G_\varepsilon(x) = F(x) + \varepsilon I(x)$ when $I(x) = x$. From theorem 2, each $G_\varepsilon(x)$ is one-one for every $\varepsilon > 0$. Note that $G_\varepsilon(x) \to F(x)$ as $\varepsilon \downarrow 0$. Hence we can conclude from More-Rheinboldt's result that $F(x)$ is one-one in R^n. This terminates the proof of theorem 2'

In fact theorem 2 can be further generalized as follows. We allow determinant $J(x)$ to vanish on a set S of isolated points in R^n.

Theorem 2"[53] : Let $n \neq 2$ and let $F:R^n \to R^n$ be continuously differentiable. Let Ω be a bounded rectangle in R^n. Suppose that the determinant $J(x)$ is positive for all x except on a set S of isolated points where it vanishes. Further suppose that every principal minor of $J(x)$ is non-negative for all $x \in R^n \setminus \Omega$. Then F is one-one.

Remark : We will not attempt to prove this result but indicate the main steps in the proof. Using (Lemma 2, pp. 244-245 in [20]) we can construct a function G having the following property (i) $\det G > 0$ for every x except on a set of isolated points and (ii) G will be norm-coercive. Now invoking a result of Chua and Lam (theorem 2.1, pp. 602-208 in [9]), we can conclude that G is one-one and consequently F is one-one. Since Chua-Lam's theorem is valid only for $n \neq 2$, we have to impose this condition in our theorem 2". For $n = 2$, Chua-Lam's theorem is not true but we do not know whether theorem 2" is true or false when $n = 2$. It is worthwhile to look at the following example which serves as a counter example in R^2 to theorem 2.1 in [9]. Let $F=(f,g)$ where $f(x,y) = x^2-y^2$ and $g(x,y) = 2xy$. Then the Jacobian is given by $J = \begin{bmatrix} 2x & -2y \\ 2y & 2x \end{bmatrix}$ and $\det J = 4(x^2+y^2)$ which vanishes only when $x = y = 0$. However F is not univalent since $F(1,1) = F(-1,-1)$. But the diagonal entries of J do change sign and as such this will not be a counter example to theorem 2".

We close this chapter by mentioning once again the following open problem: Can continuity of the derivatives in Mas-Colell's result be dispensed with or alternatively - are there counter examples ?

CHAPTER VI

GLOBAL UNIVALENT RESULTS IN R^2 AND R^3

Abstract : In this chapter we prove global univalent results obtained by Gale Nikaido and Schramm when F is a map in R^2. Some extensions are indicated in R^3. However the following two interesting open problems posed by Gale-Nikaido remain unanswered : (i) Suppose F is a differentiable map from a rectangular region $\Omega \subset R^3$ to R^3. Suppose the Jacobian is non-vanishing and every entry in the Jacobian is non-negative. Is F one-one? (ii) Suppose F is a differentiable map from a rectangular region $\Omega \subset R^3$ to R^3. Suppose every principal minor of the Jacobian is non-vanishing for every $x \in \Omega$. Is F one-one ? Gale and Nikaido have shown that both (i) and (ii) together imply that F is one-one in any rectangular region in R^3. We have shown that (i) together with the assumption that the diagonal entries are identically zero will imply that F is one-one in any open convex region in R^3-this result supplements the result obtained by Gale and Nikaido. We can weaken our assumptions in rectangular regions in R^3 using Garcia-Zangwill's result given in the previous chapter.

Univalent mappings in R^2 : There are four results which we present in this section which are due to Gale, Nikaido and Schramm. The results are rather fragmentary but are presented, for they suggest possible generalizations - in certain directions. In the first two results we relax conditions on the Jacobian matrix while in the third Ω is assumed to be a region bounded by an x-curve.

Let $F:\Omega \to R^2$ be a mapping given by $F(x,y) = (f(x,y), g(x,y))$, $(x,y) \in \Omega$ where Ω is a region in R^2. The first result concerns itself with one-signed principal minors [19].

Theorem 1 : (a) Let Ω be an arbitrary rectangular region either closed or not closed. Suppose F has continuous partial derivatives and suppose none of the principal minors of the Jacobian vanishes. Then F is univalent.

(b) Let Ω be an open rectangular region. Let the partial derivatives of F be continuous. Suppose the Jacobian does not vanish and no diagonal entries of the Jacobian matrix change signs. Then F is univalent.

Proof (a) : Since F has continuous partial derivatives, every principal minor will keep the same sign throughout Ω. That is, they will be everywhere positive or negative. We will assume without loss of generality that the diagonal entries f_x, g_y of the Jacobian matrix to be positive. In case the Jacobian is also positive then from Gale-Nikaido's fundamental theorem, F is univalent. Suppose the Jacobian is negative. This means $f_x g_y < f_y g_x$. Since $f_x, g_y > 0$ it follows f_y and g_x should be of the same sign. In case f_y and $g_x > 0$ consider the map $\tilde{F}(x,y) = (g(x,y), f(x,y))$. Then the Jacobian of $\tilde{F}$ is a P-matrix and $\tilde{F}$ is univalent in other words F is univalent. In case $f_y < 0$, $g_x < 0$ consider the map $G(x,y) = (-f(x,y), -g(x,y))$. Then the Jacobian of G is an N-matrix of the first kind. In this case also (G is univalent and hence) F is univalent. Similar proofs can be given for (b) also. This terminates the proof of theorem 1.

In the next theorem we will assume one-signedness of the entries in some row of the Jacobian matrix to assert univalence.

Theorem 2 (a) : Let Ω be an arbitrary rectangular region. Suppose F has continuous partial derivatives and the Jacobian never vanishes. If there are real numbers a and b such that $af_x + bg_x$ and $af_y + bg_y$ do not change sign and one of them does not vanish, then F is univalent.

(b) Let Ω be an open rectangular region and let F be differentiable. Suppose the Jacobian is everywhere either positive or negative. If there are real numbers a and b not both of them zero such that $af_x + bg_x$ and $af_y + bg_y$ do not change sign in Ω, then F is univalent.

Proof : (a) We will assume $a \neq 0$ for by hypothesis one of the functions $af_x + bg_x$, $af_y + bg_y$ does not vanish. Observe that the mapping $\bar{F}(x,y) = (a\,f(x,y) + bg(x,y), g(x,y))$ is univalent if and only if the original mapping F is univalent. Jacobian of $\bar{F}$ is given by

$$J_{\bar{F}} = \begin{bmatrix} af_x + bg_x & af_y + bg_y \\ & \\ g_x & g_y \end{bmatrix}$$

Then $\det J_{\tilde{F}} = a \det J_F$.

Since $af_x + bg_x$ and $af_y + bg_y$ do not change sign and one of them does not vanish and since the partial derivatives are continuous, we will assume $af_x + bg_x > 0$ and $af_y + bg_y \geq 0$. In view of this analysis we may finally assume that the original mapping F has the following properties $f_x > 0$ and $f_y \geq 0$.

Now suppose that F is not univalent, so that $F(p,q) = F(r,s) = (\alpha, \beta)$ for some distinct points (p,q), $(r,s) \in \Omega$. Clearly $q \neq s$; for if $q = s$ then $p \neq r$ and this will imply $f_x = 0$ for some point contradicting our assumption $f_x > 0$. Choose any fixed y satisfying $q \leq y \leq s$ [We assume $q < s$]. Since $f_y \geq 0$, $\alpha = f(p,q) \leq f(p,y)$, $\alpha = f(r,s) \geq f(r,y)$. By the continuity of f we have an x between p and r such that $f(x,y) = \alpha$. Since $f_x > 0$, for each y in $[q,s]$ there will exist a unique $x = \phi(y)$ such that $f(\phi(y),y) = \alpha$. Since $f(x,y)$ has continuous derivatives, ϕ is also continuously differentiable, its derivative is given by $\phi'(y) = -f_y(\phi(y),y)/f_x(\phi(y),y)$. Define $G(y) = g(\phi(y),y)$. Then G is continuously differentiable and its derivative is given by, $G'(y) = |J|/f_x$ evaluated for $x = \phi(y)$ where $|J|$ denotes the determinant of the Jacobian. Since $|J|/f_x \neq 0$, G is strictly monotonic in $[q,s]$. This contradicts the fact that $G(q) = G(s) = \beta$. Therefore F must be univalent. This terminates the proof of part (a) of theorem 2.

Proof (b) : This can be reduced to case (a) as follows. In any case we can assume $f_x \geq 0$ and $f_y \geq 0$ throughout Ω (which is assumed to be an open rectangle). Suppose $F(p,q) = F(r,s)$ for two distinct points (p,q), (r,s) belonging to Ω. Since Ω is an open set we can find real numbers α_i, β_i $(i = 1,2)$ and an open set U containing (p,q) and (r,s) such that

$$U = \{(x,y) : \alpha_1 < x < \beta_1 \quad \text{and} \quad \alpha_2 < y < \beta_2\} \quad .$$

Since $\alpha_1 < p$, $r < \beta_1$ and $\alpha_2 < q,s < \beta_2$ for a suitably chosen small positive number ε we can find an open subset U^* of U containing (p,q), (r,s) where $U^* = \{(x,y)\} : \alpha_1 < x < \beta_1,\ \alpha_2 - \varepsilon\alpha_1 < y - \varepsilon x < \beta_2 - \varepsilon\beta_1\}$. We will now define a new mapping $H:\Lambda \to R^2$ where $\Lambda = \{(x,y):\alpha_1 < x < \beta_1, \alpha_2 - \varepsilon\alpha_1 < y < \beta_2 - \varepsilon\beta_1\}$ and $H(x,y) = (h(x,y),\ k(x,y))$ with $h(x,y) = f(x,\varepsilon x+y)$ and $k(x,y) = g(x,\varepsilon x+y)$. Note that $(x,\varepsilon x+y) \in U^*$ iff $(x,y) \in \Lambda$. The Jacobian matrix J_H of H is given by

$$\begin{bmatrix} f_x(x,\varepsilon x+y) + \varepsilon f_y(x,\varepsilon x+y) & f_y(x,\varepsilon x+y) \\ g_x(x,\varepsilon x+y) + \varepsilon g_y(x,\varepsilon x+y) & g_y(x,\varepsilon x+y) \end{bmatrix} .$$

Clearly det J_H is nonvanishing. Also note that f_x and f_y both cannot be zero for det $J \neq 0$ and consequently we have $f_x(x,\varepsilon x+y) + \varepsilon f_y(x,\varepsilon x+y) > 0$ and $f_y \geq 0$. Thus $h_x > 0$ and $h_y \geq 0$ for $\forall (x,y) \in \Lambda$. Clearly the mapping H satisfies all the conditions stipulated in part (a) of theorem 2 and hence H is univalent in Λ. But Λ contains two distinct points $(p,q-\varepsilon p)$ and $(r,s-\varepsilon r)$ which are mapped to the same point under H and hence we arrive at a contradiction. Therefore F is one one in Ω and this terminates the proof of theorem 2.

Remark : It is not clear how to formulate theorem 2 in higher dimensions (when $n > 2$).

To state the next theorem we need the following [68].

Definition : An x-simple domain in the (x,y)-plane is the interior domain determined by an x-simple curve, which is a Jordan curve cut in two points at most by every line parallel to the x-axis, except possibly by lines passing through points on the curve which have extremal values of y. Analogously y-simple curves and y-simple domain are defined.

Observe that an x-simple curve can be put in the form $\bigcup_{i}^{4} \ell_i$ such that for suitable mappings $s_i:[c,d] \to R$ with $c < d$ and $s_1 < s_2$ in (c,d) we have $\ell_1=\{(s_1(y),y): c \le y \le d\}$, $\ell_2 = \{(x,c):s_1(c) \le x \le s_2(c)\}$, $\ell_3 = (s_2(y),y) : c \le y \le d\}$, $\ell_4 = \{(x,d):s_1(d) \le x \le s_2(d)\}$. The following definition helps us to identify one of the two subarcs separated by two points on a Jordan curve.

Definition : Let $P \neq Q$ be two points on a Jordan curve ℓ and λ_1, λ_2 the subarcs with $\ell \setminus \{P,Q\} = \lambda_1 \cup \lambda_2$. When λ is a subset of $\bar{\lambda}_i$ for $i = 1$ or $i = 2$ define $(P, Q, \lambda)=\lambda_i$; when μ is a proper subarc of ℓ which contains λ_i for $i = 1$ or $i = 2$ define $(\mu,P,Q)=\lambda_i$.

Definition : A square matrix over the reals is called an NVL matrix when its leading minors do not vanish. We are now ready to state and prove the following [68].

Theorem 3 : Let Ω be an x-simple domain in the (x,y)-plane, ℓ its boundary. Let $F = (f,g) : \bar{\Omega} \to R^2$ be a differentiable map, α the minimum and β the maximum of f on ℓ. Suppose the Jacobian of F is an NVL matrix for each $z \in \Omega$ and for each $u \in (\alpha,\beta)$, suppose at most two points $z \in \ell$ satisfy $f(z) = u$. Then

(a) the sign of f_x in Ω is constant

(b) defining $A(u) = \{z:z \in \bar{\Omega}$ and $f(z) = u\}$, both $A(\alpha)$ and $A(\beta)$ are subarcs of ℓ and

(c) F restricted to $\bar{\Omega} \setminus (A(\alpha) \cup A(\beta))$ is univalent.

Remarks : When conditions on univalence are arranged in the partial order of stringency of either the local conditions or the boundary conditions the above result will occupy an intermediate position - when we compare it with the Gale-Nikaido's fundamental theorem and Kestelman's result. Clearly every P-function is an NVL function and NVL functions have invertible derivatives. Contrary to a conjecture of Samuelson (see Gale-Nikaido's example) NVL functions are not unconditionally injective.

Also note that theorem 3 asserts the univalence of F in the whole of Ω, which is not necessarily a rectangular region and which need not be convex. We will now prove theorem 3.

Proof (a) : Let $c,d,s_1,s_2,\ell_1,\ell_2,\ell_3$ and ℓ_4 have the same meaning as defined above. By the Darboux property sign $f_x(.,y)$ is a function X of y only, for $c < y < d$. We will now show that the set $\{y:y \in (c,d)$ and $X(y) = 1\}$ is either empty or open in (c,d). Let y_0 be an element with $X(y_0) = 1$. This means $f_x(.,y_0) > 0$ where $y_0 \in (c,d)$. Since $s_2 > s_1$, we have $f(s_2(y_0),y_0) > f(s_1(y_0),y_0)$. As f is continuous for $|y-y_0|$ small enough $f(s_2(y),y) > f(s_1(y),y)$ and hence $f_x(.,y) > 0$. This proves (a) of theorem 3. Without loss of generality, assume f_x to be positive in Ω. Consequently $f(.,c)$ and $f(.,d)$ are non-decreasing.

We will now prove (b). From (a) we have $A(\alpha) \subset \ell$, $A(\alpha)$ meets ℓ_1 and $A(\alpha)$ meets ℓ_3 at the end points at most. Anologous assertions hold good for $A(\beta)$. To show that $A(\alpha)$ is an arc, note first that f restricted to ℓ has exactly two solutions to $f(z) = u$ when $\alpha < u < \beta$ since f-u changes sign on ℓ. [In other words if $f(z_1) = \alpha$, $f(z_2) = \beta$ where z_1, z_2 are on ℓ and since one can travel in two different directions from z_1 to z_2, $f(z) = u$ has at least two solutions on ℓ. However by hypothesis, $f(z) = u$ can have at most two solutions]. Now we shall prove that whenever $z_1 \varepsilon A(\alpha)$, $z_2 \varepsilon A(\alpha)$ the arc $\mu = (\ell_1 \cup \ell_2 \cup \ell_4, z_1, z_2)$ belongs to $A(\alpha)$. Let $0 < \varepsilon < \beta - \alpha$ and let $\lambda_\varepsilon, \lambda^\varepsilon$ be the subarcs of ℓ on which $f < \alpha + \varepsilon$, $f > \alpha + \varepsilon$, respectively. Therefore $z_1 \varepsilon \lambda_\varepsilon$, $z_2 \varepsilon \lambda_\varepsilon$ and $\mu \subset \lambda_\varepsilon$. Since ε is arbitrary $f(\mu) = \{\alpha\}$. $A(\beta)$ is treated similarly. This proves (b) of theorem 3.

We will now prove (c). Since the given region is bounded by a Jordan curve and since f_x is assumed to be positive (without loss of generality) in Ω, it is sufficient to prove the following: (i) For $\alpha < u < \beta$ we have $A(u) = \{(h(u,y),y) : y_*(u) \le y \le y^*(u)\}$ with unique functions $y_* : (\alpha,\beta) \to [c,d]$, $y^* : (\alpha,\beta) \to [c,d]$ and $h(u,.) : [y_*(u), y^*(u)] \to R$ and (ii) the functions $\alpha(u,.) : [y_*(u), y^*(u)] \to R$ defined by $\alpha(u,y) = g(h(u,y),y)$ are strictly monotone when $\alpha < u < \beta$. We will first prove (i). Note that u-points z_1, z_2 of f restricted to ℓ separate subarcs ν_1, ν_2 of ℓ with $f < u$ on ν_1, $f > u$ on ν_2. It is clear that the y coordinates of z_1 and z_2 have to be different for, first $f(.,y)$ is strictly increasing in (c,d) and second $f(.,c)$, $f(.,d)$ are nondecreasing and each has two u-points at most (by hypothesis). Let y_* and y^* be the smaller and larger of the y-coordinates of z_1, z_2 respectively. Let $\xi_1 = (s_1(y_*), y_*)$, $\xi_2 = (s_1(y^*), y^*)$, $\xi_3 = (s_2(y_*), y_*)$, $\xi_4 = (s_2(y^*), y^*)$, $\Lambda_L = \{(s_1(y),y) : y_* \le y \le y^*\}$ and $\Lambda_R = (s_2(y),y) : y_* \le y \le y^*\}$. Clearly $\Lambda_L \subset \bar{\nu}_1$, $\Lambda_R \subset \bar{\nu}_2$ as $f_x > 0$. Hence for each $y \varepsilon [y_*, y^*]$ exactly one point $(h(u,y),y) \varepsilon \bar{\Omega}$ satisfies $f(g(u,y),y) = u$. As f-u changes sign on none of (ξ_2, ξ_4, ℓ_4), (ξ_1, ξ_3, ℓ_2), no other point $(x,y) \varepsilon \bar{\Omega}$ satisfies $f = u$. To prove (ii) note that h is differentiable at (u,y) for $\alpha < u < \beta$, $y_*(u) < y < y^*(u)$. (As in the proof of theorem 2) we can conclude that $\alpha_y(u,y) = (J/f_x)_{(h(u,y),y)} \neq 0$ since Jacobian is an NVL function. This shows that $\alpha(u,.)$ is strictly monotone. This terminates the proof of theorem 3.

Remark : Theorem 3 does not ensure the univalence of F on $\bar{\Omega}$ as the following simple example shows. Let $f(x,y) = x(1-y)$ and $g(x,y) = y$ and $\bar{\Omega} = [1,2] \times [0,1]$. The mapping $F = (f,g)$ transforms $\bar{\Omega}$ into the triangle with vertices (1,0),(2,0), (0,1) the upper side of $\bar{\Omega}$ shrinking into (0,1). Also note that in the interior of the rectangle $f_x = (1-y) = |J| > 0$. We will now give an example which will satisfy the hypothesis of theorem 3. Let $\bar{\Omega} = [0,1] \times [0,1]$ with Ω interior of $\bar{\Omega}$. Let $f(x,y) = x^2 y$ and $g(x,y) = xy^2$. Clearly the map $F = (f,g)$ transforms two adjoining sides of $[0,1] \times [0,1]$ into (0,0) and satisfies the conditions of theorem 3. The following corollary is helpful in drawing conclusions about the behaviour of F on $\partial\Omega$.

Corollary : Suppose F satisfies the set of conditions of theorem 3 from which the following condition is deleted: "For each $u \in (\alpha,\beta)$ let two points $z \in \ell$ at most satisfy $f(z) = u$". Suppose F restricted to Ω is not one-one. Then for all u in some subinterval of (α,β) the equation f(restricted to ℓ) = u has more than two solutions.

Proof of the corollary follows from theorem 3.

Remark : To illuminate the corollary, let us again look at the example due to Gale and Nikaido. Let $F(x,y) = (e^{2x}-y^2+3,\ 4e^{2x}y-y^3)$ over $\bar{\Omega} = [0,1] \times [-2,2]$. Here $\ell_1 = \{(0,y):-2 \leq y \leq 2\}$, $\ell_3 = \{(1,y):-2 \leq y \leq 2\}$, $\ell_2 = \{(x,c):0 \leq x \leq 1, c = -2\}$ and $\ell_4 = \{(x,d):0 \leq x \leq 1,\ d=2\}$. It turns out both $F(\ell_1)$ and $F(\ell_2 \cup \ell_3 \cup \ell_4)$ are Jordan curves symmetrical with respect to the x-axis. Also $F(\ell_1)$ lies in the interior determined by $F(\ell_2 \cup \ell_3 \cup \ell_4)$ except for the origin which belongs to both the curves. Thus for each u in a certain interval the equation $f|\ell = u$ has more than two solutions.

Theorem 4 : In addition to conditions of theorem 3, suppose that for both $y_0 = c$ and $y_0 = d$ either $s_1(y_0) = s_2(y_0)$ or $f(.,y_0)$ has a nonvanishing derivative in $(s_1(y_0),\ s_2(y_0))$ and F is NVL on $\ell \setminus \ell_2 \setminus \ell_4$. Then F is one-one on $\bar{\Omega}$ (= interior plus the boundary).

Proof : In view of theorem 3(c) it is enough to prove that F is one one on both $A(\alpha)$ and $A(\beta)$. We will prove theorem 4 when $s_1(c) = s_2(c)$, $s_1(d) = s_2(d)$. A similar proof can be given in the other case. From theorem 3(b) we can conclude that $A(\alpha) \subset \ell_1$ or $A(\alpha) = \{(s_1(y),y) : y_*(\alpha) \leq y \leq y^*(\alpha)\}$ for some $y_*(\alpha)$ and $y^*(\alpha) \in R$. Assume $y_*(\alpha) < y^*(\alpha)$ and define $\gamma(\alpha,y) = g(s_1(y),y)$. In order to complete the proof of the theorem it is enough if we show that $\gamma_y(\alpha,.) \neq 0$ in $(y_*(\alpha),y^*(\alpha))$. Let $y_0 \in (y_*(\alpha),\ y^*(\alpha))$ and let f and g have partial derivatives (P,Q) and (M,N) at $z_o = (s_1(y_0),y_0)$. From the local implicit function theorem it follows that $s_1'(y_0)$ exists and it is given by $s_1'(y_0) = -\frac{Q}{P}$. Also $\gamma_y(\alpha,y_0) = (\frac{PN-QM}{P}) \neq 0$ as F is NVL at z_o. This terminates the proof of theorem 4.

(Note that $F'(z_o)$ is unique since $P = f_x, M = g_x$, $Q = -Ps_1'$ and $N = ((P\gamma_y+QM)/P)$.

We will now present a result due to Nikaido [45]. Let $F = (f(x,y),g(x,y))$ be a mapping from R^2 to R^2 where f and g have continuous partial derivatives throughout R^2.

Theorem 5 : Suppose there are 4 positive numbers m_1, m_2 and M_1, M_2 such that

$$m_1 \leq |f_x| \leq M_1$$

and

$$m_2 \leq |f_x g_y - f_y g_x| \leq M_2$$

throughout R^2. Then the system of equations

$$f(x,y) = a_1$$

$$g(x,y) = a_2$$

has exactly one solution in R^2 for any given constants a_1, a_2. In other words F is one-one and onto R^2.

Proof : Observe that the existence and uniqueness of a solution of a single equation $f(x) = a$ in the single unknown x are clearly true if $m \leq |f'(x)| \leq M(-\infty < x < \infty)$ for some positive m and M. If this condition is satisfied, one can see from the mean value theorem in calculus that either $\lim_{x \to \pm\infty} f(x) = \pm \infty$ or $\lim_{x \to \pm\infty} f(x) = \mp \infty$ holds true, depending on the sign of the derivative $f'(x)$. Hence, $f(x)$ can be equated to a at some value of x because of the continuity of $f(x)$. Moreover the solution is unique from the montonicity of $f(x)$.

Suppose the conditions of theorem are met. This means given a_1 and $y, f(x,y) = a_1$ has a unique solution from the first paragraph of the proof. That is there exists a function $\phi(y) = x$, such that $f(\phi(y),y) = a_1$. Also one can check by the local implicit function theorem that ϕ has continuous derivative ϕ' with respect to y which will satisfy, $f_x(\phi(y),y)\phi'(y) + f_y(\phi(y),y) = 0$. Let $G(y) = g(\phi(y),y)$. Then $G'(y) = g_x(\phi(y),y)\phi'(y) + g_y(\phi(y),y)$. Hence it follows that $G'(y) = - g_x \frac{f_y}{f_x} + g_y$

From the hypothesis, $\frac{m_2}{M_1} \leq |G'(y)| \leq \frac{M_2}{m_1}$. Again from the first paragraph it follows that $G(y) = a_2$ has unique solution for any given constant a_2. This shows that F is one-one and onto. This terminates the proof of this theorem.

This result can be generalized to higher dimensions and the proof goes through with very few changes.

Theorem 6 : Let $F = (f_1, f_2, \ldots, f_n)$ be a mapping from R^n to R^n, with continuous partial derivatives $f_{ij} = \frac{\partial f_i}{\partial x_j}$ in the whole of R^n. Suppose that there are 2n positive numbers m_k and $M_k, 1 \leq k \leq n$ such that the absolute values of the upper left-hand corner principal minor determinants satisfy

$$m_k \leq \text{absolute value of} \begin{vmatrix} f_{11} & \cdots & f_{1k} \\ & & \\ f_{k1} & \cdots & f_{kk} \end{vmatrix} \leq M_k \quad (k = 1,2,\ldots,n)$$

in the whole of R^n. Then the system of equations

$$f_i(x_1,x_2,\ldots,x_n) = a_i \quad (i = 1,2,\ldots,n)$$

has exactly one solution in R^n for any given constants a_i. In other words F is one-one and onto and consequently F is a homeomorphism of R^n onto R^n.

Counter example : We will now present a counter example to illustrate that the truth of theorem 5 depends very heavily on the assumption that the domain of F is the whole of R^2. Consider the system of equations

$$f(x,y) = x^2-y^2+3 = 0$$

$$g(x,y) = 4x^2y-y^3 = 0 .$$

Let $\Omega = \{(x,y) : \frac{1}{2} \leq x \leq 2, -3 \leq y \leq 3\}$. Also note that

$$J = \begin{bmatrix} 2x & -2y \\ & \\ 8xy & -3y^2+4x^2 \end{bmatrix} , \quad 4 > |2x| > 1 \text{ and}$$

$244 > |8x^3 + 10xy^2| > 1$ in the region Ω. But

$$f(1,2) = g(1,2) = 0 = f(1,-2) = g(1,-2).$$

This example is due to Nikaido, (and Gale-Nikaido's example given elsewhere is a slight modification of this example). Theorem 5 is similar in nature to that of theorem 3 in the sense that we impose conditions on the leading minors and not on every principal minor of the Jacobian matrix. It is not clear whether theorem 6 can be deduced from the results presented in chapter IV.

We will now pose an interesting open problem in R^2 on univalent mappings. Let F be a continuously differentiable map from R^2 to R^2. Suppose (i) trace $J(x) < 0$ and (ii) det $J(x) > 0$ for every $x \in R^2$. Is F univalent ? In other words if for every x, $J(x)$ has only characteristic roots with negative real parts, is F one-one? Note that none of the results proved here covers this interesting case. However the result is not true if (i) trace $J(x) < 0$ and (ii) det $J(x) < 0$ as the following example shows. Let $F(x,y) = (f,g)$ when $f(x,y) = -2e^x + 3y^2-1$ and $g(x,y) = ye^x-y^3$. Then it is easy to see that $|J| = -2e^{2x} < 0$ and trace $J = -e^x-3y^2 < 0$ and F maps (0,1) as well as (0,-1) to the same point (0,0).

<u>Univalent results in R^3</u> : The following result is due to Gale and Nikaido though a proof of which is never published. The proof which we give here is incomplete and we invite the readers to furnish a complete proof.

<u>Theorem 7</u> : Let $F:\Omega \subset R^3 \to R^3$ be a $C^{(1)}$ map where Ω is a compact rectangular region. Suppose every entry in the Jacobian matrix is positive throughout Ω. Further suppose that $\det J(x) > 0$ and all other principal minors are non-vanishing. Then F is one-one.

It is clear that principal minors of order one are positive. If every principal minor of order 2 is positive, then the Jacobian matrix is a P-matrix throughout Ω and consequently F is one-one. If every principal minor of order 2 is negative then the Jacobian of the map -F is an N-matrix and once again from Gale-Nikaido's fundamental theorem -F is one-one or F is one-one. The difficult case appears to be the following. Suppose the leading principal minor of order 2 is negative and the other principal minors are positive. In this case we are unable to show that F is one-one. However we can prove the following result.

<u>Theorem 8</u> : Let $F:\Omega \subset R^3 \to R^3$ be a $C^{(1)}$ map where Ω is a compact rectangular region. Let $\det J > 0$ for every $x \in \Omega$. Further suppose diagonal entries of the Jacobian are 0 and the off-diagonal entries are positive throughout $\partial\Omega$. Then F is one-one in Ω.

<u>Proof</u> : Let $F = (f,g,h)$. Then the Jacobian of F is given by

$$J = \begin{bmatrix} 0 & f_y & f_z \\ g_x & 0 & g_z \\ h_x & h_y & 0 \end{bmatrix}$$

where all the partial derivatives are positive by hypothesis on $\partial\Omega$. Let $G = (g,h,f)$. Then clearly Jacobian G is non-vanishing throughout Ω and it is a P-matrix on $\partial\Omega$. Hence Garcia-Zangwill theorem will imply that G is one-one or F is one-one. This terminates the proof of theorem 8.

The following theorem complements theorem 7 of Gale-Nikaido in R^3 [53].

<u>Theorem 9</u> : Let $F : \Omega \subset R^3 \to R^3$ be a $C^{(1)}$ map where Ω is an open rectangular region with determinant of the Jacobian positive throughout Ω. Further suppose diagonal entries of the Jacobian are zero and the off-diagonal entries are non-negative throught Ω. Then F is one-one in Ω.

<u>Proof</u> : Let $F = (f,g,h)$ where (f,g,h) are real-valued maps from Ω. Let G_ε be a map from Ω to R^3 defined as follows:

$$G_\varepsilon(x,y,z) = (g(x,y,z) + \varepsilon x, h(x,y,z) + \varepsilon y,\ f(x,y,z) + \varepsilon z)$$

where ε is any positive number. Then it is easy to check that the Jacobian of the map G_ε is a P-matrix for every $\varepsilon > 0$ and further $G_\varepsilon \to (g,h,f)$ for every $(x,y,z) \in \Omega$. Hence F is univalent in Ω. This terminates the proof of theorem 9.

In fact theorem 9 holds good in any open convex region.

Theorem 10 : Let $F:\Omega \subset R^3 \to R^3$ be a $C^{(1)}$ map where Ω is an open convex region in R^3 with $\det J > 0$ throughout Ω. Further suppose the diagonal entries are zero and the off-diagonal entries are non-negative throughout Ω. Then F is univalent in Ω.

Proof : For every $\varepsilon > 0$, let G_ε be the map defined as follows: $G_\varepsilon(x,y,z) = (f+\varepsilon z, g+\varepsilon x, h+\varepsilon y)$. Then the Jacobian of G_ε is given by

$$\begin{bmatrix} 0 & f_y & f_z+\varepsilon \\ g_x+\varepsilon & 0 & g_z \\ h_x & h_y+\varepsilon & 0 \end{bmatrix} .$$

Now one can use argument similar to proof of theorem 4 in chapter III to conclude that G_ε is one-one in Ω. Now an application of More-Rheinboldt's result yields the fact that F is univalent in Ω. This terminates the proof of theorem 10.

Remark : It is not clear whether any of the univalent results proved above remain ture in R^4. We are unable to get any counter example in R^4 for theorems 9 and 10. In R^3 the most challenging problems appear to be the following: (1) Suppose F is a differentiable map from a rectangular region $\Omega \subset R^3$ to R^3. Suppose the Jacobian is non-vanishing and every entry is non-negative throughout Ω. Is F one-one in Ω ? (2) Suppose F is a differentiable map from a rectangular region $\Omega \subset R^3$ to R^3. Suppose every principal minor of the Jacobian is non-vanishing throughout Ω. Is F one-one in Ω ? (3) Suppose F is a differentiable map from a convex region $\Omega \subset R^3$ to R^3. Suppose every entry in the Jacobian is non-negative and every principal minor is non-vanishing throughout Ω. Is F one-one in Ω? Note that in R^2 all these results hold good.

In R^2 the following problem appears to be unresolved. Let $\Omega \subset R^2$ be a bounded and closed rectangle whose sides are not parallel to the axes. Suppose the Jacobian J of a continuously differentiable map $F:\Omega \to R^2$ is positive throughout Ω and on the boundary of Ω it is a P-matrix. Is F one-one in Ω ? This result is true if the Jacobian is a positive definite or quasi-positive definite matrix on the boundary of Ω. It is also true if F is one-one on the boundary of Ω.

CHAPTER VII

ON THE GLOBAL STABILITY OF AN AUTONOMOUS SYSTEM ON THE PLANE

Abstract : In this chapter we consider an old problem of Olech (which is equivalent to a global univalent problem in R^2) on the global stability of an autonomous system on the plane and extend a result of Olech using some of the recent global univalent theorems (see [52]). We present also a variety of examples to show the sharpness of the results. In the second half of this chapter we present the results obtained by Vidossich [71] in this direction.

Introduction : Consider an autonomous system (S) ... $x' = F(x)$, (Here prime denotes derivative with respect to time point t) where $x = (x_1,x_2)$, $F(x) = (f_1(x_1,x_2), f_2(x_1,x_2))$. Suppose $x = 0 = (0,0)$ is a critical point of (S) and assume that the Jacobian matrix J at each point x of R^2 has characteristic roots with negative real parts. That is assume (i) determinant of $J > 0$ and (ii) trace of $J < 0$ for every $x \in R^2$. [Of course we do assume F to be a continuously differentiable function throughout R^2]. Is then the solution $x = 0$ of (S) global asymptotically stable, or in other words does each solution curve of (S) approach the critical point 0 as $t \to \infty$? It is not very hard to see that this problem is equivalent to the following global univalent problem [47].

Global Univalent Problem : Let $F:R^2 \to R^2$ be a continuously differentiable function. Suppose the Jacobian matrix J of F has the following two properties for every x in R^2: (i) det $J > 0$ and (ii) trace of $J < 0$. Is then F globally univalent or one-one throughout R^2.

It is not known whether global univalent problem has an affirmative answer. If it has, then one can conclude that the solution $x = (0,0)$ of (S) is globally asymptotically stable. However in [47] it is shown that $x = 0$ is globally asymptotically stable if (S) satisfies (i), (ii) and if F is a one-one map. We have the following theorem due to Hartman, Markus and Yamabe and Olech [25, 34, 47].

Theorem 1 : Let the autonomous system (S) satisfy (i) and (ii). Let $F(0) = 0$. Then the solution $x = 0$ is globally asymptotically stable if any one of the following conditions is met.

[MY] one of the four partial derivative $\frac{\partial f_i}{\partial x_j}$ ($i = 1,2$, and $j = 1,2$) vanishes identically on R^2.

[H] if the symmetric part of the Jacobian matrix is negative definite on R^2.

[0] Either $\left(\frac{\partial f_1}{\partial x_1}\right)\left(\frac{\partial f_2}{\partial x_2}\right) \neq 0$ on R^2 or $\left(\frac{\partial f_1}{\partial x_2}\right)\left(\frac{\partial f_2}{\partial x_1}\right) \neq 0$ on R^2.

Remark : Condition (0) is automatically satisfied if either [MY] or [H] is satisfied. We now have the following theorem which includes theorem 1 [52].

Theorem 2 : Let the autonomous system satisfy (i) and (ii) and $F(0) = (0,0)$ and let F be a continuously differentiable function throughout R^2. Write $F = (f,g)$. Suppose any one of the following sets is bounded in R^2.

(a) $D = \{(x,y) : \frac{\partial f}{\partial x} \cdot \frac{\partial g}{\partial y} < 0\}$

(b) $E = \{(x,y) : \frac{\partial f}{\partial y} \cdot \frac{\partial g}{\partial x} < 0\}$

(c) $G = \{(x,y) : \frac{\partial f}{\partial x} \cdot \frac{\partial g}{\partial y} > 0\}$.

Then the solution $(x,y) = (0,0)$ is globally asymptotically stable. (Note that we have slightly altered the notation; instead of writing f_1, f_2 we have written (f,g) and instead of writing (x_1,x_2) we have written (x,y)).

Proof of theorem 2 uses results from Olech [47] More and Rheinboldt [42] and Garcia-Zangwill [20]. Then we will give several examples to illustrate the sharpness of the results. In particular we will give an example where theorem 2 is applicable but not theorem 1.

Proof of Theorem 2 : We will give a proof of theorem 2. In view of Theorem 3, (pp. 395 in [47]) it is enough if we show that the mapping F is globally one-one under each of the assumptions (a), (b) and (c). We will write f_x, f_y instead of $\frac{\partial f}{\partial x}, \frac{\partial f}{\partial y}$ etc. Suppose (a) holds. That is we are given that the set $D=[(x,y):f_x g_y<0]$ is bounded. We will assume without loss of generality that D (as well as E and G) are compact rectangles. We can always take the closure of the set D if necessary. Note that $f_x\, g_y \geq 0$ throughout $R^2 \setminus D$. We will now consider two cases (α) $f_x g_y > 0$ throughout $R^2 \setminus D$ and (β) $f_x g_y = 0$ for some point in $R^2 \setminus D$. If (α) holds good, since $f_x + g_y < 0$ by hypothesis it follows that $f_x < 0$ and $g_y < 0$ throughout $R^2 \setminus D$. Invoking theorem 3 (pp. 246 in [20]) we may conclude that $-F = (-f, -g)$ is globally univalent or that F is globally univalent in R^2.

If (β) holds good, then we define $F_\varepsilon(x,y)$ for every $\varepsilon > 0$ as follows.

$$F_\varepsilon(x,y) = F(x,y) - (\varepsilon x, \varepsilon y).$$

Observe that the J_ε = Jacobian of $F_\varepsilon = \begin{bmatrix} f_x-\varepsilon & f_y \\ g_x & g_y-\varepsilon \end{bmatrix}$. Now det $J_\varepsilon = f_xg_y + \varepsilon^2 - \varepsilon(f_x+g_y) - g_xf_y > 0$ for every $\varepsilon > 0$ since $f_x + g_y < 0$ and det $J > 0$. Also trace of $J_\varepsilon = f_x + g_y - 2\varepsilon < 0$. Furthermore product of the diagonal entries of J_ε is given by

$$(f_x-\varepsilon)(g_y-\varepsilon) = f_xg_y - \varepsilon(f_x+g_y) + \varepsilon^2$$

which is strictly positive throughout $R^2 \setminus D$. In other words $F_\varepsilon(x,y)$ satisfies all the conditions of theorem 3 of Garcia-Zangwill [20] and consequently $F_\varepsilon(x,y)$ is globally univalent in R^2 for every $\varepsilon > 0$. That is $-F_\varepsilon(x,y) = -F(x,y) + (\varepsilon x,\varepsilon y)$ is globally univalent in R^2 for every $\varepsilon > 0$. Also Jacobian of $-F$ is non-vanishing and R^2 is an open set. Thus we may infer from theorem 5.9 of More and Rheinboldt [42] that $-F(x,y)$ or $F(x,y)$ is globally univalent in R^2. This terminates the proof under condition (a).

Suppose (b) holds good. If the set E is bounded, since the Jacobian of F is positive it follows that $D \subset E$ or the set D is bounded and we are in case (a).

Suppose (c) holds good. Since $f_xg_y \leq 0$ throughout $R^2 \setminus G$ and since det $J > 0$, we have $f_y\, g_x < 0$ throughout $R^2 \setminus G$. As F is of $C^{(1)}$ class and $R^2 \setminus G$ is a connected set, it follows that f_y and g_x will keep the same sign throughout $R^2 \setminus G$. Suppose $f_y < 0$ and $g_x > 0$ throughout $R^2 \setminus G$. Let $H = (g,-f)$. Then the Jacobian J_H of H is a P-matrix - (that is every principal minor of J_H is positive) throughout $R^2 \setminus G$. Hence from Garcia-Zangwill's result it follows that H is globally univalent or F is one-one throughout R^2. A similar proof can be given when $f_y > 0$ and $g_x < 0$ throughout $R^2 \setminus G$. Thus we see that each one of the conditions (a), (b) and (c) imply that the map F is univalent in R^2 and consequently $(x,y) = (0,0)$ is globally asymptotically stable. This terminates the proof of theorem 2.

We will now deduce theorem 1 from theorem 2. In order to do that it is enough to verify that condition [0] implies one of the conditions (a), (b) and (c) of theorem 2. If $f_x\, g_y < 0$ throughout R^2, condition (c) is trivially satisfied as G is an empty set. If $f_xg_y > 0$ throughout R^2, condition (a) is met as D is an empty set. If $f_yg_x > 0$ throughout R^2 it will imply $f_xg_y > 0$ is R^2 and D will be an empty set. Suppose $f_yg_x < 0$ throughout R^2. We will consider two cases (1) $f_y < 0$ and $g_x > 0$ in R^2 and (2) $f_y > 0$ and $g_x < 0$. Under case (1) define a new map $H = (-g,f)$. Then condition (a) is met for the map H. Under case (2) define a new map $L=(g,-f)$. Then again condition (a) is met for L. We may conclude from theorem 2, $(x,y)=(0,0)$ is globally asymptotically stable.

One has the following corollary which can be deduced from theorem 2.

Corollary 1 : Let the autonomous system satisfy (i) and (ii) and $F(0) = (0,0)$. Let $F = (f,g)$ be a $C^{(1)}$ function. Suppose $f_x g_y \geq 0$ throughout R^2. Then the solution $(x,y) = (0,0)$ is globally asymptotically stable.

Remark : Clearly corollary 1 includes Markus and Yamabe's result as well as Hartman's result. Also Olech's condition given in theorem 1 need not be satisfied. We will utilize this corollary to construct an example where theorem 1 may not be applicable.

Examples: In this section we will give several examples to illustrate the limitations or the sharpness of the results known for the global asymptotic stability of the solutions for the autonomous system in the plane. In the first two examples Olech's condition given in theorem 1 will not be satisfied but conditions given in theorem 2 will be satisfied. Fourth example will illustrate that condition (ii) trace of $J < 0$ cannot be weakened to trace of $J \leq 0$, to get the asymptotic stability results. If condition (i) is replaced by (i') det $J < 0$ throughout R^2 and retain condition (ii) then one can have more than one critical points - we will demonstrate this in the third example. The next two examples will show that it is possible to have det $J = 0$ at some points and yet $(0,0)$ to be a globally asymptotic solution. In other words condition (i) or the map F to be one-one need not be necessary. In these two examples we will use the following result due to Hartman and Olech (Theorem 2.1, pp. 155 in [27]).

Theorem 3 : Let F be a continuously differentiable function from R^2 to R^2. Suppose $F(0) = 0$ and $F(x) \neq 0$ whenever $x \neq 0$. Suppose $x = 0$ is a locally asymptotic stable solution of (S). Let $\lambda_1(x)$ and $\lambda_2(x)$ be the characteristic roots of the Jacobian of F evaluated at x. Suppose $\lambda_1(x) + \lambda_2(x) \leq 0$ for all $x \in R^2$ and further suppose $|x|\ |f(x)| \geq$ constant > 0 whenever $|x| \geq$ constant > 0. Then $x = 0$ is globally asymptotically stable.

Our last example will illustrate that theorem 3 may fail if $\lambda_1(x) + \lambda_2(x)$ admits positive values.

Example 1 : Let $f_1(x_1,x_2) = -x_1 + \log(1+x_1^2) + x_2$ and $f_2(x_1,x_2) = -x_1^3 - e^{x_2} + 1$. Here

$$J = \begin{bmatrix} -1 + \left(\dfrac{2x_1}{1+x_1^2}\right) & 1 \\ \\ -3x_1^2 & -e^{x_2} \end{bmatrix} .$$

Clearly det $J > 0$ and trace $J < 0$. Note that $\dfrac{\partial f_1}{\partial x_1} \cdot \dfrac{\partial f_2}{\partial x_2} = 0$ if $x_1 = 1$ and

$\frac{\partial f_1}{\partial x_2} \cdot \frac{\partial f_2}{\partial x_1} = 0$ if $x_1 = 0$. Also $\frac{\partial f_1}{\partial x_1} \cdot \frac{\partial f_2}{\partial x_2} \geq 0$ throughout R^2. Here all the conditions of corollary 1 are satisfied (but Olech's condition given in theorem 1 is not satisfied). Consequently $x = 0$ is globally asymptotically stable.

Example 2 : Let $f_1(x_1,x_2) = -\left[\frac{x_1^3}{3} + x_1(x_2^2 - \tfrac{1}{2}) - x_2\right]$

$$f_2(x_1,x_2) = -[x_1+x_2].$$

Here the Jacobian J is given by

$$J = \begin{bmatrix} -(x_1^2+x_2^2-\tfrac{1}{2}) & -2x_1x_2+1 \\ -1 & -1 \end{bmatrix}.$$

Clearly $\det J > 0$, trace $J < 0$. In this example the set D given in theorem 2, is a bounded set. In fact $D = \{(x_1,x_2) : x_1^2 + x_2^2 \leq \tfrac{1}{2}\}$. Here also theorem 1 is not applicable. We can conclude from theorem 2 that $x = 0$ is globally asymptotically stable.

Example 3 : This example is a slight modification of Gale-Nikaido given in (pp 82, [19]). Let

$$f_1(x_1,x_2) = -2e^{x_1} + 3x_2^2 - 1$$

$$f_2(x_1,x_2) = x_2 e^{x_1} - x_2^3.$$

$$J = \begin{bmatrix} -2e^{x_1} & 6x_2 \\ x_2e^{x_1} & e^{x_1}-3x_2^2 \end{bmatrix}$$

Clearly $\det J < 0$ and trace $J < 0$. However $F = (f_1,f_2)$ is not globally univalent since $F(0,1) = F(0,-1) = (0,0)$. This example shows that global univalent problem is false if condition (i) is replaced by the condition $\det J < 0$ throughout R^2. As already remarked global univalent problem still remains open. A modification of the same example will serve as a counter example if one attempts to prove global univalent problem in R^3 under (i) and (ii).

Example 4 : If we weaken condition (ii) to trace $J \leq 0$ then global asymptotic stability may not hold good as the following example shows. Set $f_1(x_1,x_2) = -x_1+2x_2$ and $f_2(x_1,x_2) = -x_1+x_2$. Here

$$J = \begin{bmatrix} -1 & 2 \\ -1 & 1 \end{bmatrix}$$

Clearly det J > 0 and trace J $\leq$ 0 throughout R^2. However here x = 0 is not even locally asymptotically stable though it is stable (see theorem 9.1, pp. 411 in [4]).

Example 5 : The following example will show that condition (i) namely det J > 0 is not a necessary condition for x = 0 to be globally asymptotically stable. Let

$$f_1(x_1,x_2) = -x_1(1+x_2^2) \quad \text{and}$$

$$f_2(x_1,x_2) = -x_2(1+x_1^2).$$

$$J = \begin{bmatrix} -(1+x_2^2) & -2x_1x_2 \\ -2x_1x_2 & -(1+x_1^2) \end{bmatrix}$$

Here det J = $(1+x_1^2)(1+x_2^2) - 4x_1^2x_2^2 = 0$ if $x_1 \pm 1$ and $x_2 = \pm 1$. Consequently condition (i) is violated. It is not hard to check that x = 0 is locally asymptotically stable. (see pp. 440 in [4]). Also one can easily check that $\lambda_1 + \lambda_2 = -(2+x_1^2+x_2^2) < 0$. Hence all the conditions of theorem 3 are satisfied and thus x = 0 is globally asymptotically stable. Alternatively one can use theorem 9.5 in pp. 439, [4] to demonstrate that x = 0 is globally asymptotically stable.

Example 6 : The following example shows that x = 0 may be globally asymptotically stable but that F need not be globally one-one in R^2. Let

$$f_1(x_1,x_2) = -\frac{x_1^3}{2} + 2x_1x_2^2$$

$$f_2(x_1,x_2) = -x_2^3$$

$$J = \begin{bmatrix} -\frac{3}{2}x_1^2 + 2x_2^2 & 4x_1x_2 \\ 0 & -3x_2^2 \end{bmatrix}$$

$\det J = 3x_2^2(2x_2^2 - \frac{3}{2}x_1^2) \neq 0$ if $x_2 \neq 0$ and $2x_2^2 \neq \frac{3}{2}x_1^2$ and trace $J \leq 0$. Also $F = (f_1, f_2)$ is not one-one since $F(1,\frac{1}{2}) = F(-1,\frac{1}{2}) = (0,-\frac{1}{2})$. One can check that $||F(x)|| \geq$ constant > 0 whenever $||x|| \geq$ constant > 0. All the conditions of theorem 3 are satisfied and thus $x = 0$ is globally asymptotically stable. [One can check that $x = 0$ is locally asymptotically stable by Liapounou's second method for details see pp. 440 in (4)].

Example 7 : The following example demonstrates that theorem 3 is probably the best available result in R^2 for global asymptotic solution. Let $\varepsilon < 0$ and $\sigma < 0$. Let

$$f_1(x_1,x_2) = \varepsilon x_1 - \sigma x_1^2$$

$$f_2(x_1,x_2) = -x_2 .$$

Then the Jacobian

$$J = \begin{bmatrix} \varepsilon - 2\sigma x_1 & 0 \\ 0 & -1 \end{bmatrix} .$$

In this example $x = 0$ is locally asymptotically stable but not globally asymptotically stable. Also note that $\lambda_1(x) + \lambda_2(x) = -1 + \varepsilon - 2\sigma x_1$, may take both positive and negative values. In other words one of the conditions of theorem 3 is not satisfied. In fact other conditions are met. This shows that in general we may not be able to relax the condition "$\lambda_1 + \lambda_2 \leq 0$" if we want globally asymptotic solution.

Vidossich's contribution to Olech's problem on stability : Vidossich's gives another set of sufficient conditions under which F is one-one. In fact we have the following [71].

Theorem 4 : Let $F:R^n \to R^n$ be of class $C^{(1)}$ and all eigenvalues of the Jacobian $J(x)$ of F have negative real parts, then F is one-one if any of the following conditions is met:

(a) there exist two positive constants α and β such that trace $J(x) \leq -\alpha$ and $||J(x)|| \leq \beta$ for every $x \in R^n$ or alternatively,

(b) for each $y \in R^n$, there exists $\varepsilon_y > 0$ such that for the topological degree we have $\deg(F,B(0,\varepsilon),y) = 1$ for $\varepsilon > \varepsilon_y$, $B(0,\varepsilon)$ being the open ball with centre 0 and radius ε.

Remarks : Theorem 4 is true for any n. When (a) is satisfied one can show that F is proper which will in turn imply that F is one-one - this can be seen from Hadamard's theorem or from Caccioppoli [6]. Also condition (b) is satisfied if there exists $\varepsilon_o > 0$ and $0 < k < 1$ such that

$$||x-F(x)|| \leq k||x||, \quad (||x|| \geq \varepsilon_o).$$

This fact follows from the homotopy invariance of the topological degree by considering $I-\lambda(I-F)$, $(0 \leq \lambda \leq 1)$.

Proof of theorem 4 : We will prove under condition (b). Suppose F is not one-one. This means for some $y_o \in R^n$, $F^{-1}(y_o)$ will contain at least two points say u,v. Since F is locally one-one and R^n is separable, $F^{-1}(y_o)$ contains at most countably many points. There exists $\varepsilon > 0$ such that

$$\varepsilon > \max\{\varepsilon_{y_o}, ||u||, ||v||\}$$

and no points of $F^{-1}(y_o)$ has norm equal to ε, since otherwise $F^{-1}(y_o)$ would contain a subset with the same cardinality of a non-trivial interval of R contrary to the countability of $F^{-1}(y_o)$. Since the closure of $B(0,\varepsilon)$ is compact and $F^{-1}(y_o) \cap B(0,\varepsilon)$ is discrete it follows that $F^{-1}(y_o) \cap B(0,\varepsilon)$ must be finite : let $x_1,\ldots,x_m$ be its points. Clearly $||x_i|| < \varepsilon$ for $i = 1,2,\ldots,m$. Hence we can find for each i an open set U_i of x_i such that the closed sets $\bar{U}_i$ are pairwise disjoint and $U_i \subset B(0,\varepsilon)$. Let

$$G(x) = x + F(x) - y_o .$$

Then, $F(x) = y_o$ if and only if $G(x) = x$. Observe that $G(x)$ has no fixed point in $B(0,\varepsilon) \setminus \bigcup_{i=1}^{m} U_i$. By the additivity of the topological degree, we have

$$\deg(I-G, B(0,\varepsilon),0) = \sum_{i=1}^{m} \deg(I-G, U_i,0)$$

where the right hand side is well defined since no point of $F^{-1}(y_o)$ belongs to ∂U_i. We will now show that $\deg(I-G,U_i,0) = 1$, $i = 1,2,\ldots,m$. Let λ be an eigenvalue of $J_G(x_i)$ = Jacobian of G. Then there exists $x \neq 0$ such that $(I+J_F(x_i))x = \lambda x$ or $(\lambda-1)x = J_F(x_i)x$. It follows that $(\lambda-1)$ is an eigenvalue of $J_F(x_i)$ and by hypothesis $\lambda - 1 < 0$ or $\lambda < 1$. It implies that J_G has no positive eigenvalue greater than one. Therefore a well-known theorem of Leray-Schauder (See pp 162-163 [48]) implies that $\deg(I-G, U_i,0) = 1$ for $i = 1,2,\ldots,m$. It is known that

$$\deg(g,B(0,\varepsilon),y) = \deg(g-y, B(0,\varepsilon),0)$$

for any continuous function g and any point y provided that the equation $g(x) = y$ has no solution x with $||x|| = \varepsilon$. Therefore we have :

$$\begin{aligned} \deg(F,B(0,\varepsilon),y_0) &= \deg(F-y_0,\ B(0,\varepsilon),0) \\ &= \deg(I-G,\ B(0,\varepsilon),0) \\ &= \sum_{i=1}^{m} \deg(I-G,\ U_i,0) \\ &= m\ . \end{aligned}$$

But $m \geq 2$, since $u,v \in F^{-1}(y_0) \cap B(0,\varepsilon)$, hence we have a contradiction to our assumption (b). This terminates the proof of theorem 4.

We now have the following theorem on stability.

Theorem 5 : The solution $x = 0$ of the autonomous equation $x' = F(x)$ in the plane is globally asymptotically stable if F is of class $C^{(1)}$, $F(0) = 0$, the eigenvalues of $J(x)$ all have negative real parts and F satisfies (a) or (b) of theorem 4.

Proof : Follows from theorem 4 and Olech's theorem.

Remark 1 : Example 6 shows that the solution $x = 0$ may be globally asymptotic stable, without any of the conditions imposed by theorem 4 being satisfied.

Remark 2 : Global asymptotic stability in R^n has been studied by Hartman [25] and Hartman and Olech [27]. Hartman considers the case when $(J+ J')/2$ is negative definite while Hartman and Olech places conditions on the eigenvalues of the Jacobian matrix. Interested readers should refer to their works for further details.

We close this chapter by inviting the readers to prove the global univalent problem or give a counter example.

CHAPTER VIII

UNIVALENCE FOR MAPPINGS WITH LEONTIEF TYPE JACOBIANS

Abstract : In this chapter we prove several results on univalence for mappings with Leontief type Jacobians. The first result is in some sense a sort of converse to Gale-Nikaido's theorem on univalence. Here we prove a result due to Gale-Nikaido and this says that if F and F^{-1} are differentiable and if F^{-1} is monotonic increasing then the Jacobian of F is a P-matrix provided the Jacobian matrix of F is of Leontief type. The second result due to Nikaido says that there exists a unique solution to $F(x) = 0$ provided its domain is non-negative orthant and the Jacobian matrix is of Leontief type satisfying certain uniform diagonal dominance property. Then we present related results on M-functions and inverse isotone maps due to More and Rheinboldt. Finally we give some results on the univalence of the composition of maps F and G when their Jacobians are of Leontief type. In particular we show that F o G is a P-function when F and G are maps from R^3 to R^3 with their Jacobians Leontief type P-matrices throughout. We give an example to show that F o G need not be a P-function in R^4.

Univalence for dominant diagonal mappings : Two results will be presented in this section. The first result shows that if both F and F^{-1} are differentiable and if F^{-1} is monotonic increasing then the Jacobian matrix of F has to be a P-matrix, provided the Jacobian matrix of F is of Leontief type. This result is due to Gale and Nikaido. The second result due to Nikaido says that there exists a unique solution to $F = 0$ provided the domain of F is Ω = non-negative orthant and the Jacobian matrix is of Leontief type satisfying certain uniform diagonal dominance property. We will say that a matrix A is of Leontief type if the off-diagonal entries of A are nonpositive. Such matrices play a prominent role in studies initiated by well-known economist Leontief. For Leontief type matrices we have seen that the following two conditions are equivalent.

(i) A is a P-matrix

(ii) There exists a vector $x > 0$ such that $Ax > 0$.

In chapter II we have defined the notion of dominant diagonal. That is, a matrix A is said to possess a dominant diagonal if there is a set of positive numbers $d_i (i = 1,2,\ldots,n)$ such that $a_{ii}d_i > \sum_{\substack{j=1 \\ j \neq i}}^{n} |a_{ij}| d_j$ for all $i = 1,2,\ldots,n$. This condition is equivalent to the condition that there exists a set of positive numbers

c_j $(j = 1,2,\ldots,n)$ such that

$$a_{jj}\, c_j > \sum_{\substack{i=1\\ i\neq j}}^{n} |a_{ij}| c_i \quad \text{for all} \quad j = 1,2,\ldots,n.$$

Equivalence follows from the fact that any one of the above conditions implies that A is a P matrix.

If A, whose elements are functions defined on a set Ω has dominant diagonal throughout the set Ω, the weights d_i in general will be a function of $w \in \Omega$. If $d_i(w) \equiv d_i$ for all $w \in \Omega$ then we say A has a uniformly dominant diagonal. We say A has a uniformly dominant diagonal in the strong sense if there are positive numbers d_i and c_i, $i = 1,2,\ldots,n$ such that $dA \geq c$ where $d = (d_1,d_2,\ldots,d_n)$, A is an $n \times n$ matrix and $c = (c_1,c_2,\ldots,c_n)$. [Entries of A are functions of w]. We are now ready to state the following theorems.

Theorem 1 : Let $F : \Omega \to R^n$ be a differentiable mapping where Ω is an open region in R^n and Jacobian matrix is of Leontief type. Suppose F^{-1} is differentiable and monotonic increasing (that is $F(a) \leq F(b)$ implies $a \leq b$). Then the Jacobian matrix of F is a P-matrix.

Theorem 2 : Let $F:R^n_+ \to R^n$ be a differentiable map whose Jacobian matrix is of Leontief type. (Here R^n_+ = non-negative orthant). Suppose Jacobian matrix of F has a uniformly dominant diagonal in the strong sense. Further suppose $f_i(x) \leq 0$ whenever the i-th component of x is zero. (Here $f_i(x)$ is the i-th component of $F(x)$). Then $F(x) = 0$ has a unique solution. In other words $F(x) = a$ has a unique solution for every $a \in R^n_+$.

Proof of Theorem 1 : Since Ω is open in R^n, by the invariance of domain, $F(\Omega)$ is also an open set in R^n. Let b be any point of Ω such that $b^* = F(b)$. Let u^* be any strictly positive vector. Since $F(\Omega)$ is open, there exists $\varepsilon > 0$ such that $x^*(\lambda) = b^* + \lambda u^* \in F(\Omega)$ for all λ with $|\lambda| < \varepsilon$. Let $F^{-1}(x^*(\lambda)) = x(\lambda)$. Then since F^{-1} is differentiable and monotonic increasing, $x(\lambda)$ is differentiable and $\frac{d}{d\lambda}x(\lambda) \geq 0$ for $|\lambda| < \varepsilon$. Differentiating $x^*(\lambda) = F(x(\lambda))$ at $\lambda = 0$, we have $J(b)x'(0) = u^* > 0$. Since $J(b)$ is of Leontief type, and since $x'(0) \geq 0$ and $u^* > 0$ it follows that $J(b)$ has to be a P-matrix. (Here as usual $J(b)$ stands for the Jacobian matrix evaluated at b). This terminates the proof of theorem 1.

Proof of theorem 2 : Proof of this result is similar to the proof of theorem 5 in chapter VI. Proof is based on induction on n, which will establish both the existence and uniqueness of the solution of the system of equations $F(x) = 0$. If $n = 1$, we are looking for a solution of a single equation of a scalar variable $f_1(x) = 0$.

We are given two positive constants d_1 and c_1 such that $d_1\ f_{11}(x) > c_1$ for all $x \geq 0$. This implies the first derivative $f_{11}(x)$ of $f_1(x)$ is positive and in fact $f_{11}(x) > (c_1/d_1)$. Hence it follows that $f_1(x)$ is strictly monotonic increasing and $f_1(x) \to +\infty$ as $x \to \infty$. Since from hypothesis $f_1(0) \leq 0$, it follows that $f_1(x) = 0$ has a solution as f_1 is continuous and the solution is unique as f_1 is strictly monotonic.

Assume the result to be true for n-1. Let us write $F(x) = 0$ explicitly as follows:

$$f_i(x_1, x_2, \ldots, x_{n-1}, x_n) = 0 \quad \text{for} \quad i = 1,2,\ldots,n-1$$

$$f_n(x_1, x_2, \ldots, x_{n-1}, x_n) = 0$$

Fix x_n and consider the system of equations in the (n-1) variables $x_1, x_2, \ldots, x_{n-1}$. The system $f_i(x_1, \ldots, x_{n-1}, x_n) = 0$ for $i = 1,2,\ldots,n-1$ satisfies all the hypothesis of the theorem including the strong uniform dominant diagonal property. To this end, let J_{n-1} denote principal submatrix of order n-1 in the upper left corner of that of the original system. Since $(d_1, \ldots, d_n)J \geq (c_1, \ldots, c_n)$ for some $d \geq 0, c \geq 0$ and since the off-diagonal entries of J are non-positive it follows that

$$(d_1, \ldots, d_{n-1})J_{n-1} \geq (c_1, \ldots, c_{n-1}).$$

Hence by induction hypothesis the system of equations $f_i(x_1, \ldots, x_{n-1}, x_n) = 0$ for $i = 1,2,\ldots,n-1$, for every fixed non-negative x_n, has a unique solution, $x_1, \ldots, x_{n-1}$ which will be functions of x_n. Let $x_j = g_j(x_n)$, $j = 1,2,\ldots,n-1$ which of course will satisfy

$$f_i(g_1(x_n), g_2(x_n), \ldots, g_{n-1}(x_n), x_n) = 0, i = 1,2,\ldots,n-1$$

for all $x_n \geq 0$. Let $g(x_n) = f_n(g_1(x_n), g_2(x_n), \ldots, g_{n-1}(x_n), x_n)$. Now it is easy to conclude that $F(x_1, \ldots, x_n) = 0$ has a unique solution if and only if $g(x_n) = 0$ is uniquely solvable in the non-negative unknown x_n. Thus we have reduced the problem to one-variable case. We will now show that $g(0) \leq 0$ and $d_n \frac{dg}{dx_n} \geq c_n$ - these two facts will complete the proof.

Obviously $g(0) = f_n(g_1(0), \ldots, g_{n-1}(0), 0) \leq 0$. Observe

$$dJ \geq c \quad \text{and} \quad J \text{ is of Leontief type.}$$

Hence it follows that J^{-1} is non-negative. Thus

$$(d_1, d_2, \ldots, d_n)J \geq (c_1, \ldots, c_n) \text{ or } (d_1, \ldots, d_n) \geq (c_1, \ldots, c_n)J^{-1}$$

Now $d_n \geq \sum_{i=1}^{n} c_i f^{in}$ where (f^{in}) is the n-th column of J^{-1}. Since

$$f^{in} \geq 0 \quad \text{and} \quad c_i \geq 0,\ d_n \geq c_n\ f^{nn} = c_n \frac{\det J_{n-1}}{\det J}$$

or

$$d_n \frac{\det J}{\det J_{n-1}} \geq c_n .$$

Note that

$$\frac{dg}{dx_n} = \frac{\det J}{\det J_{n-1}} .$$

Hence $g(x_n) = 0$ will have a unique solution and consequently $F(x) = 0$ will have a unique solution. This terminates the proof of theorem 2.

Interrelation between P-property and M-property : We will now introduce the notion of M-function similar to the notion of P-function introduced in chapter III and prove that if F is a differentiable map over the rectangle and if the Jacobian is a P-matrix of the Leontief type then F must be an M-function.

Definition : Consider a mapping $F:\Omega \subset R^n \to R^n$ with the following properties

(a) $F(x) \leq F(y) \Rightarrow x \leq y$ whenever $x,y \in \Omega$

(b) the functions $\Psi_{ij}:\{t \in R^1 | x+ te^j \in \Omega\} \to R^1$ defined by $\Psi_{ij}(t) = f_i(x+te^j)$ are monotonic decreasing as functions of t where $i \neq j$ and $e^1 = (1,0,0,\ldots,0)$, $e^2 = (0,1,0,\ldots,0)$ etc. [Here f_i as usual denotes the ith component of F].

Then F is said to be an M-function. Property (a) is usually called inverse isotone and (b) is called off-diagonally antitone. This notion is a nonlinear generalization of Leontief type P-matrices introduced and studied by More and Rheinboldt. We are ready to prove the following theorem.

Theorem 3 : Let $F:\Omega \subset R^n \to R^n$ be a differentiable map on the rectangle Ω. Suppose the Jacobian matrix of F is a P-matrix of the Leontief type for every $x \in \Omega$. Then F is an M-function, in the sense described above.

Remark : It is known from Gale-Nikaido's result that F is a P-function. Also note that F is off-diagonally antitone. Theorem 3 is an immediate consequence of the following result.

Theorem 4 : Any off-diagonally antitone P-function $F:\Omega \subset R^n \to R^n$ on a rectangle Ω is an M-function.

Proof : We need only to verify property (a) of the above definition of an M-function. That is, we have to establish that F is inverse isotone. Suppose $F(x) \leq F(y)$ for some $x,y \in \Omega$. Let $A = \{i:x_i > y_i\}$. If A is empty we are through. Suppose A is not empty and let $A = \{1,2,\ldots,m\}$. Define

$$g_i(t_1,t_2,\ldots,t_m) = f_i(t_1,t_2,\ldots,t_m,y_{m+1},\ldots,y_n)$$

where $i = 1,2,\ldots,m$. Note that

$$g_i(y_1,y_2,\ldots,y_m) = f_i(y) \geq f_i(x) \geq g_i(x_1,x_2,\ldots,x_m).$$

Here the last inequality follows from the fact that F is off-diagonally antitone and the first inequality by hypothesis. Hence we can conclude that

$$(y_i-x_i)(g_i(y_1,\ldots,y_m)-g_i(x_1,\ldots,x_m)) \leq 0, \quad i = 1,2,\ldots,m.$$

Set $x' = (x_1,x_2,\ldots,x_m, y_{m+1},\ldots,y_n)$. Clearly $x' \neq y$ and

$$(y_i-x_i')(f_i(y) - f_i(x')) \leq 0, i = 1,2,\ldots,m$$

and

$$(y_i-x_i')(f_i(y) - f_i(x')) = 0, i = m+1,\ldots,n$$

but this contradicts the fact that F is a P-function. This terminates the proof of theorem 4. Now theorem 3 follows from theorem 4.

Remark : We need the fact that Ω is a rectangle to conclude that $x' \in \Omega$. It is not clear whether this result is valid if Ω is a convex region and not necessarily a rectangular region. We have lot of freedom when Ω is a rectangular region. One can restate theorem 1 as follows, for convex maps in terms of property (a).

Theorem 5 : Let $F:\Omega \subset R^n \to R^n$ be convex and differentiable on the open convex set Ω. Then F is inverse isotone if and only if the Jacobian matrix $J(x)$ is invertible for each $x \in \Omega$ and $J(x)^{-1} \geq 0$ for each $x \in \Omega$ (that is every entry in the inverse matrix is non-negative).

Proof : Suppose F is inverse isotone. That is $F(x) \leq F(y)$ implies $x \leq y$. Let $x \in \Omega$ and suppose $J(x)h \geq 0$. Since Ω is open, there is a $\theta > 0$, such that $x+ \theta h \in \Omega$. It follows from the convexity of F

$$F(x+\theta h)-F(x) \geq \theta J(x)h \geq 0 \quad .$$

The last inequality follows because $\theta > 0$ and $J(x)h \geq 0$ by assumption. Since F is inverse isotone, $x+ \theta h \geq x$ or $h \geq 0$. In other words $J(x)h \geq 0$ implies $h \geq 0$. Consequently it follows that $J(x)$ is non-singular. Now we will prove $J(x)^{-1} \geq 0$. In order to do that, let $y \geq 0$ then we have $J(x)(J(x)^{-1} y) \geq 0$ which in turn implies $J(x)^{-1} y \geq 0$. Hence $J(x)^{-1} \geq 0$.

Conversely suppose $J(x)^{-1} \geq 0$ for every $x \in \Omega$. Let $F(x) \geq F(y)$ for some $x,y \in \Omega$ From the convexity of F it follows that $J(x)(x-y) \geq F(x)-F(y) \geq 0$. That is

$J(x)(x-y) \geq 0$ or $J(x)^{-1} J(x)(x-y) = x-y \geq 0$. This terminates the proof of theorem 5.

Remark : If in addition we assume $J(x)$ is of Leontief type in theorem 5 then F is inverse isotone if and only if $J(x)$ is a P-matrix for each $x \in \Omega$. Inverse isotone concept is equivalent to fact that the mapping F^{-1} is monotonic increasing provided F^{-1} exists. If we drop the assumption of convexity of F as well as convexity of Ω we have the following theorem.

Theorem 6 : Let $F:\Omega \subset R^n \to R^n$ be inverse isotone and differentiable on the open set Ω.Suppose the Jacobian $J(x)$ of F is of Leontief type. Then $J(x)^{-1}$ is a P-matrix for any $x \in \Omega$ at which $J(x)$ is non-singular.

Proof : Suppose $x \in \Omega$ with $J(x)$ non-singular. In order to show $J(x)$ is a P-matrix it is enough if we show $J(x)^{-1} \geq 0$ since $J(x)$ is of Leontief type. As in theorem 5 we will prove the following $J(x)h \geq 0$ for some $h \in R^n$ will imply $h \geq 0$. This can be seen as follows. First we will consider the case $J(x)h > 0$.

$$\lim_{t \to 0^+} [F(x+th)-F(x)] = J(x)h > 0 .$$

Hence $F(x+th)-F(x) > 0$ for sufficiently small $t > 0$ and consequently by the inverse isotonicity that $h \geq 0$. Now consider $J(x)h \geq 0$ for some $h \in R^n$. Define $h^k = h + \frac{1}{k} J(x)^{-1} e$, $k = 1,2,\ldots$. Here e denotes the vector with all entires one. Clearly $J(x)h^k > 0$ which in turn implies from the first part $h^k \geq 0$, for all k and hence $h \geq 0$ since $h^k \to h$ as $k \to \infty$. This shows $J(x)^{-1} \geq 0$. This terminates the proof of theorem 6. Observe that theorem 6 is essentially the same as theorem 1. For open rectangular regions one can prove the following theorem.

Theorem 7 : Let $F:\Omega \subset R^n \to R^n$ be a differentiable map on the open rectangular set Ω. Suppose $J(x)$ is of Leontief type and suppose $J(x)$ is non-singular for each $x \in \Omega$. Then F is inverse isotone if and only if $J(x)$ is a P-matrix for every $x \in \Omega$.

Proof : If F is inverse isotone then from theorem 6, it follows that $J(x)$ is a P-matrix for every $x \in \Omega$. Now suppose $J(x)$ is a P-matrix. From Gale-Nikaido's fundamental theorem F is univalent in Ω. We will prove that F is inverse isotone by induction argument on n. When $n = 1$, clearly F is inverse isotone. In general, if $F(a) \leq F(b)$ where $a = (a_1,a_2,\ldots,a_n)$ and $b = (b_1,b_2,\ldots,b_n)$, then for some k we have $a_k \leq b_k$ this is a consequence of theorem 1, chapter III. We may assume without loss of generality $k = 1$ that is $a_1 \leq b_1$. Since the off-diagonal entries of the Jacobian matrix are non-positive, we have for $i > 1$,

$$f_i(b_1,a_2,\ldots,a_n) \leq f_i(a_1,a_2,\ldots,a_n) \leq f_i(b_1,b_2,\ldots,b_n).$$

Here the second inequality follows from $F(a) \leq F(b)$. We now define a new map $G:\Omega_{n-1} \subset R^{n-1} \to R^{n-1}$ by the rule $G(x_2,x_3,\ldots,x_n) = (f_2(b_1,x_2,\ldots,x_n),\ldots,f_n(b_1,x_2,\ldots,x_n))$ where Ω_{n-1} is the image of Ω under the projection $(x_1,x_2,\ldots,x_n) \to (x_2,\ldots,x_n)$. Now $G(a_2,\ldots,a_n) \leq G(b_2,\ldots,b_n)$ and the Jacobian matrix of G is again a P-matrix of the Leontief type. Hence by induction hypothesis $a_i \leq b_i$ for $i > 2$. We already have $a_1 \leq b_1$ and hence $a_i \leq b_i$ for every i. This proves F is inverse isotone and this terminates the proof of Theorem 7.

Univalence for composition of two functions : In this section we will prove in R^3, that if F and G are differentiable functions whose Jacobian matrices are Leontief type P-matrices then F o G is a P-function and hence univalent in R^3. However one can prove the following result in R^n. If F and G are differentiable functions and if Jacobian matrices of F and G are Leontief type P-matrices and further if the Jacobian of F o G is also of Leontief type then F o G is a P-function and hence univalent in R^n. The results we prove in this section are similar to theorem 3.4 in More and Rheinboldt [42].

We are ready to prove the following [53].

Theorem 8 : Let F and G be $C^{(1)}$ differentiable maps from $R^n \to R^n$. Suppose Jacobians of F and G are Leontief type P-matrices. Further suppose Jacobian of the composition map F o G is of Leontief type. Then F o G is a P-function and hence univalent in R^n.

Proof : Let H = F o G. Observe that the Jacobian J_H of H can be computed by means of the relation $J_H = J_F\, J_G$ where the elements of J_F as well as J_G will be evaluated at the appropriate points. From theorem 4, chapter II we can conclude that J_H is a P-matrix throughout R^n. Invoking Gale-Nikaido's theorem, we have the desired result.

Theorem 9 : Let F and G be $C^{(1)}$ differentiable maps from R^3 to R^3. Suppose Jacobians of F and G are Leontief type P-matrices. Then F o G and G o F are P-functions and hence univalent in R^3.

Proof : We will prove F o G is a P-function. Let H = F o G. From theorem 4, chapter II, we can conclude that the Jacobian of H is a P-matrix and consequently F o G is a P-function. Similar proof may be given for the univalence of G o F.

From theorem 2, chapter V we can deduce the following:

Theorem 10 : Let F and G be $C^{(1)}$ differentiable maps from R^n to R^n. Let Ω be a compact rectangle in R^n. Further suppose Jacobians of F and G are P-matrices in $R^n \setminus \Omega$ respectively. Then F o G is univalent in R^n.

We gave an example in chapter II (see theorem 4) to show that the product of two Leontief type P-matrices in R^4 need not be either a P-matrix or a Leontief type matrix. That example will correspond to the following functions in R^4:

$$F(x,y,z,w) \quad = \quad (x-z, \ \tfrac{1}{8} y-w, \ z, \ w)$$

$$G(x,y,z,w) \quad = \quad (x-y, \ -x+ 2y, \ -4y+ z, \ -x+w).$$

Then

$$(F \circ G)(x,y,z,w) \quad = \quad (x+ 3y-z, \tfrac{7}{8} x + \tfrac{2y}{8} - w, \ -4y+ z, \ -x+ w).$$

In this example Jacobian of F and G are Leontief type P-matrices. However Jacobian of F o G is a non-singular (constant) matrix which is not a P-matrix. Also it is easy to see in this example that F o G is one-one throughout R^4. This raises the following question: Suppose F and G are differentiable $C^{(1)}$ function in R^4 whose Jacobian matrices are Leontief type P-matrices. Is F o G a P-function in R^4? The answer is "no" and the example given above is a counter example and it can be seen as follows. Jacobian matrix of F o G is a constant matrix A given by

$$A = \begin{bmatrix} 1 & 3 & -1 & 0 \\ \frac{7}{8} & \frac{2}{8} & 0 & -1 \\ 0 & -4 & 1 & 0 \\ -1 & 0 & 0 & 1 \end{bmatrix}$$

Since A is not a P-matrix, then exists a non-trivial vector $u^o = (x^o, y^o, z^o, w^o)$. such that $(Au^o)'_i \, u^o_i \leq 0$ for $i = 1,2,3,4$ [Here prime denotes the transpose vector and u^o_i stands for the ith component of the vector u^o]. Note that $(F \circ G)(u^o) = Au^{o'}$ Let θ be the zero vector. Clearly $u^o \neq \theta$ and $(F \circ G)(\theta) = 0$. Hence it follows that F o G is not a P-function.

We will now deviate a little bit and start with two examples. The first is an example of a differentiable P-function whose Jacobian is not a P-matrix. Consider $F:R^2 \to R^2$ where $F = (f,g)$ with $f(x,y) = x^3-y$ and $g(x,y) = x+y^3$. One can easily check that F is a P-function throughout R^2. However Jacobian J of F given by

$$J = \begin{bmatrix} 3x^2 & -1 \\ 1 & 3y^2 \end{bmatrix} = 1 + 9x^2y^2 > 0$$

is not a P-matrix at $x = y = 0$.

The second example is the following.

This says that if the Jacobian is a weak P-matrix in a rectangular region, F need not be a P-function. Consider $F = (f,g)$ where $f(x,y) = y$, $g(x,y) = y-x$. Here $J = \begin{bmatrix} 0 & 1 \\ -1 & 1 \end{bmatrix}$ is a weak P-matrix but F is not a P-function. We know from Gale-Nikaido's fundamental theorem that this result is true if J is a P-matrix.

We will now introduce the concept of P_0 function which is weaker than P-function

<u>Definition</u> : A mapping $F:\Omega \subset R^n \to R^n$ is a P_0-function if for any $x,y \in \Omega$, $x \neq y$, there is an index $k = k(x,y)$ such that $(x_k-y_k)(f_k(x)-f_k(y)) \geq 0$ and $x_k \neq y_k$.

For P_0-functions one can prove the following theorem [42].

<u>Theorem 11</u> : Let $F:\Omega \subset R^n \to R^n$ be a differentiable P_0-function with $\det J(x) > 0$ for every $x \in \Omega$ where Ω is an open rectangle. Then F^{-1} is again a P_0-function.

<u>Remark 1</u> : We will not attempt to give a proof but we will indicate the proof. Let D be any diagonal matrix with entires non-negative. Then one can check that $F(x) + Dx$ is a P_0-function which will in turn imply $J(x) + D$ is a weak P-matrix. In other words $F(x) + Dx$ is univalent for every such D. This in turn implies that F^{-1} is a P_0-function.

<u>Remark 2</u> : A differentiable P_0-function need not possess a weak P-matrix as its Jacobian. It will only be a P_0-matrix, that is, every principal minor is non-negative.

We will close this chapter with a conjecture of More-Rheinboldt [42]: For any continuous, injective P_0-function $F:\Omega \subset R^n \to R^n$ on an open rectangle Ω, F^{-1} is also a P_0-function. This result certainly holds good for the linear case and for F-differentiable P_0-functions. In order to settle this conjecture it is enough if one can show that $F(x) + Dx$ is one-one for every diagonal matrix D with non-negative entries.

CHAPTER IX

ASSORTED APPLICATIONS OF UNIVALENCE MAPPING RESULTS

Abstract : In this chapter we will give various applications of the univalence results proved in the earlier chapters. There are several areas where univalence results are quite handy and useful. The first application will deal with a problem in Mathematical Economics where we will give a set of sufficient conditions due to Nikaido and Mas-Colell which will ensure factor price equalization. The second application deals with the distribution of a function of several independent random variables. As a third application we will consider a problem in nonlinear complimentarity theory due to Kojima and Megiddo. Next we give an application of Hadamard's theorem to Algebra. In the fifth we consider the problem of deciding whether a certain multivariate gamma distribution is infinitely divisible. In this situation weak N-matrices play an important role. There are various other applications (for example to nonlinear net-work theory) but we will not attempt to exhaust all of them for lack of time and space. [We have already seen a nice application of univalent results in stability theory in chapter VII].

An application in Mathematical Economics : As a first application we will consider a well known problem in mathematical economics first suggested by Samuelson which in a sense prompted Gale-Nikaido to prove their fundamental result on univalence mappings. In this problem one seeks conditions implying that countries facing the same prices for goods in foreign trade will have the same factor prices. To be precise we will give a set of sufficient conditions which will not only ensure factor price equalization but also a set of equilibrium factor prices under any given set of good prices. Put differently, the condition ensures the complete invertibility of the determination of good prices by factor prices, giving rise to the inverse unique determination of factor prices by arbitrary good prices.

We will now explain the terminology that will be used. Let $w_j > 0$, $p_i > 0$ denote the price of the jth factor and the ith good respectively where $j,i=1,2,\dots,n$. Let $c_i(w_1,w_2,\dots,w_n)$ be n cost functions which have continuous non-negative partial derivatives $c_{ij} = \partial c_i/\partial w_j$ and c_i's are positively homogeneous of degree one, that is $c_i(\lambda w_1,\dots,\lambda w_n) = \lambda c_i(w_1,\dots,w_n)$ for all $\lambda > 0$ and $w_j > 0$. Let $\alpha_{ij} = c_{ij}\, w_j/c_i$ - this quantity is the relative share of the jth factor in the ith good sector, because $\sum_{j=1}^{n} c_{ij}\, w_j = c_i$. [The last equality follows from the fact that c_i's are positively homogeneous of degree one]. By 'equalization of factor prices' we mean the uniqueness of solution of the system of equations $c_i(w_1,w_2,\ldots,w_n) = p_i$ $(i = 1,2,\ldots,n)$. We are ready to state the complete invertibility theorem due to Nikaido [45].

Complete invertibility theorem : Suppose $c_i(w_1,w_2,\ldots,w_n)$ are defined for all $w_j > 0$, take positive values everywhere, have continuous partial derivatives c_{ij} which are non-negative everywhere and c_i's are positively homogeneous of degree one. Further suppose the relative shares matrix $A = (\alpha_{ij})$ where $\alpha_{ij} = c_{ij}\, w_j/c_i$ has the upper left-hand corner principal minors whose absolute values are bounded from below by some number

$$\text{absolute value of det} \begin{bmatrix} \alpha_{11} & \cdots & \alpha_{1k} \\ \vdots & & \\ \alpha_{k1} & \cdots & \alpha_{kk} \end{bmatrix} \geq \delta, (k = 1,2,\ldots,n) .$$

Then for any given set of positive good prices $p_i > 0$, there exists a unique set of factor prices $w_j > 0$ satisfying $c_i(w_1,w_2,\ldots,w_n) = p_i$ for $i = 1,2,\ldots,n$.

Proof : Set $f_i = \log c_i$ and $w_j = e^{x_j}$. In other words we are defining n new functions defined on R^n

$$f_i(x_1,x_2,\ldots,x_n) = \log c_i(e^{x_1}, e^{x_2},\ldots,e^{x_n}) .$$

It is clear that the system of equations $c_i(w_1,\ldots,w_n) = p_i$ has a unique solution for $p_i > 0$ if and only if the system of equations $f_i(x_1,x_2,\ldots,x_n) = a_i$ where $a_i = \log p_i$ has a unique solution in R^n. We will now verify that f_i's satisfy the conditions imposed on theorem 6 in chapter VI. Note that

$$\frac{\partial f_i}{\partial x_j} = e^{x_j}\, c_{ij}(e^{x_1},e^{x_2},\ldots,e^{x_n})/c_i(e^{x_1},\ldots,e^{x_n})$$

so that $f_{ij} = \alpha_{ij}$ if evaluated for $w_j = e^{x_j}$ in R^n. Hence δ given in the theorem will serve as a lower bound. Also observe by homogeneity, $\sum_j \alpha_{ij} = 1$ for $c_i = \Sigma\, c_{ij}\, w_j$ and further by hypothesis $\alpha_{ij} \geq 0$. Hence $0 \leq \alpha_{ij} \leq 1$ and consequently the principal minors of (f_{ij}) are bounded above. Therefore the system $f_i = a_i$ has a unique solution for any given $p_i > 0$ from theorem 6 chapter VI. This terminates the proof of the complete invertibility theorem.

Remark : Mas-Colell has substantially generalized the complete invertibility theorem. He shows that the restrictions on the principal minors are irrelevant; all that matters is that that the determinant of $|\alpha_{ij}|_{nxn}$ be uniformly bounded away from zero. For details see Mas-Colell [36].

We will now look at another example from input-output model recently considered by Chander. Chander's model is similar to the model considered earlier by I.W. Sandberg [1973, Econometrica pp. 1167-1182]. We assume that the economy is divided into n industrial sectors each of which produces a single kind of good that is traded, consumed and invested in the economy. The interrelations in such an economy may be described by the system of equations

$$x_i - \sum_{j=1}^{n} a_{ij}(x_j) = c_i \quad \text{for} \quad i = 1,2,\ldots,n$$

where x_i denotes the quantity of good i produced in the ith sector, $a_{ij}(x_j)$ represents the total amount of good i used as input for producing x_j units of good j. Hence, for each i the total amount of good i available for final consumption, export and investment is $x_i - \sum_{j=1}^{n} a_{ij}(x_j)$. The vector $(c_1,c_2,\ldots,c_n)$ is called the final demand vector. Here the problem is to find the vector x given the demand vector c. Also one would be interested in computing x. We will make two assumptions: (i) for each i and j, $a_{ij}(.)$ is continuously differentiable on $[0,\infty)$, $a_{ij}(0) = 0$ and $a'_{ij}(\alpha) \geq 0$ for all $\alpha \geq 0$ where prime denotes the derivative. (ii) There exists $p_i > 0$, $i = 1,2,\ldots,n$ and $v > 0$ such that $p_j \geq \sum_{i=1}^{n} p_i \, a'_{ij}(\alpha) + v$ for all $\alpha \in [0,\infty)$ and $j=1,2,\ldots,n$. We now have the following:

Theorem : (for input-output model) : Under assumptions (i) and (ii) the mapping $A : R^n \to R^n$ where $A(x) = (\sum_{j=1}^{n} a_{ij}(x_j),\ i = 1,2,\ldots,n)$ is a contraction mapping. Furthermore for every $c \in R^n_+$, there exists a unique $x \in R^n_+$ such that $x-A(x) = c$ and for any $x(0) \in R^n_+$, the sequence $\{x(t)\}_0^\infty$ defined by the Jacobi iterates

$$x(t+1) = A(x(t)) + c, \quad t \geq 0.$$

converges to x.

For a proof of this result one can refer to Chander [7]. Alternatively one can use Nikaido's result on uniform diagonal property in the strong sense. Here we are making stronger assumption and this enables us to compute the vector x to any degree of accuracy that we want.

On the distribution of a function of several random variables : Usually in statistics we come across problems of the following type. If $X_1,X_2,\ldots,X_n$ are mutually independent random variables then we would like to know the distribution of a function $u(X_1,X_2,\ldots,X_n)$ of the random variables $X_1,X_2,\ldots,X_n$ or we may even be interested in the joint distribution of $F(X_1,X_2,\ldots,X_n) = (u_1(X_1,X_2,\ldots,X_n), u_2(X_1,X_2,\ldots,X_n)\ldots u_n(X_1,\ldots,X_n))$. To fix ideas and show how univalence results are useful in such a situation we will work out a simple example. Let X_1,X_2,X_3 be mutually independent random variables each having a gamma distribution with $\beta = 1$. Then the joint distribution of X_1, X_2, X_3 is given by

$$\phi(x_1,x_2,x_3) = \prod_{i=1}^{3} \frac{1}{\Gamma(\alpha_i)} x_i^{\alpha_i-1} e^{-x_i}, \quad 0 \le x_i \le \infty \text{ and } \alpha_i > 0$$

$$= 0 \qquad \text{otherwise.}$$

Let

$$u_1(x_1,x_2,x_3) = x_1/(x_1+x_2+x_3)$$

$$u_2(x_1,x_2,x_3) = x_2/(x_1+x_2+x_3)$$

and

$$u_3(x_1,x_2,x_3) = x_1 + x_2 + x_3 \;.$$

Problem is to find the joint distribution of (U_1, U_2, U_3) (we are using capital letters to denote random variables). In order to do that, first we have to check whether the map $F = (u_1, u_2, u_3)$ is one-one in the effective domain of (x_1, x_2, x_3). In this example we have to verify whether F is one-one in the positive orthant of R^3. We now write down the Jacobian matrix

$$J = \begin{bmatrix} \frac{u_3-x_1}{u_3^2} & -\frac{x_1}{u_3^2} & -\frac{x_1}{u_3^2} \\ \frac{x_2}{u_3^2} & \frac{u_3-x_2}{u_3^2} & -\frac{x_2}{u_3^2} \\ 1 & 1 & 1 \end{bmatrix}$$

One can easily check that the Jacobian matrix is a P matrix for every x in the positive orthant of R^3. In fact in this example one can easily write down the inverse map. In fact $F^{-1}(u_1,u_2,u_3) = (u_1u_3,\ u_2u_3,\ u_3(1-u_1-u_2))$ where (u_1,u_2,u_3) is a point in the range of F. Now one can write down the joint distribution of U_1, U_2, U_3 by means of the following formula.

$$\Psi(u_1,u_2,u_3) = \frac{1}{|J|} \phi(u_1u_3,\ u_2u_3,\ u_3(1-u_1-u_2))$$

whenever $u_1 > 0$, $u_2 > 0$, $u_1+u_2 < 1$ and $0 < u_3 < \infty$ and $\Psi = 0$ otherwise. Explicit expression for $\Psi(u_1,u_2,u_3)$ is given by

$$u_2^{\alpha_2-1}\ u_3^{\alpha_1+\alpha_2+\alpha_3-1}\ u_1^{\alpha_1-1}\ (1-u_1-u_2)^{\alpha_3-1}\ e^{-u_3}/(\Gamma(\alpha_1)\Gamma(\alpha_2)\Gamma(\alpha_3))$$

In particular joint distribution h of (U_1,U_2) is given by

$$h(u_1,u_2) = \frac{\Gamma(\alpha_1+\alpha_2+\alpha_3)}{\Gamma(\alpha_1)\Gamma(\alpha_2)\Gamma(\alpha_3)} u_1^{\alpha_1-1} u_2^{\alpha_2-1} (1-u_1-u_2)^{\alpha_3-1}$$

when $0 < u_1,u_2$ and $u_1 + u_2 < 1$ and $h = 0$ otherwise. Random variables U_1,U_2 that have a joint distribution of this form are said to have a Dirichlet distribution with parameters α_1, α_2, α_3. One can easily check that the marginal of U_1 or U_2 is a beta distribution.

On the existence and uniqueness of solutions in Nonlinear Complimentarity theory:

Let $F : R^n_+ \to R^n$ be a continuous mapping where R^+_n is the non-negative orthant of R_n. We are interested in finding a non-negative vector $z \in R^n_+$ such that $F(z) \in R^n_+$ and $z.F(z) = 0$. Here $z.F(z)$ stands for the inner product between the two vectors z and $F(z)$. We call this problem the complimentarity problem associated with F. The CP (= Complimentarity Problem) associated with F is said to be globally uniquely solvable if for any vector $q \in R^n$ the CP associated with $F(\cdot) + q$ has a unique solution. We will follow the approach of Megiddo and Kojima to give a solution to the problem under consideration. Let G be an extension of F defined as follows. G is a map from R^n to R^n. Define $G(x) = F(x^+) + x^-$ where

$$x_i^+ = \begin{cases} x_i & \text{if } x_i \geq 0 \\ 0 & \text{if } x_i < 0 \end{cases} \quad \text{and} \quad x^- = \begin{cases} x_i & \text{if } x_i \leq 0 \\ 0 & \text{if } x_i > 0 \end{cases}$$

Let $F:R^n_+ \to R_n$ be a continuous mapping. Then the CP associated with F is globally uniquely solvable if and only if for every $q \in R^n$, there is a unique $x = x(q) \in R^n_+$ such that $F(x) + q \in R^n_+$ and $f_i(x) + q_i = 0$ for those i where $x_i > 0$. This then is equivalent to the existence of a unique $z = z(q) \in R^n$ such that $F(z^+) + q = - z^-$ or or $G(z) = - q$. Now we have the following.

Theorem 1 : Let $F:R^n_+ \to R^n$ be a continuous mapping. Then the CP associated with F is globally uniquely solvable if and only if extension G of F is a homeomorphism of R^n onto itself.

Remark : Suppose $F:R^n_+ \to R^n$ is a continuous mapping such that the CP associated with F is globally uniquely solvable. Then the solution of the CP associated with $F(.)+ q$ is a continuous function of the vector q for every $q \in R^n$. This is a consequence of the fact that, if $z = G^{-1}(q)$ then $x = z^+$ is a solution for the CP associated with $F(\cdot) + q$. Now we shall prove the following theorem giving sufficient conditions on the map F such that the CP associated with F+ q will have at most one solution for every $q \in R^n$.

Theorem 2 : (a) If $F:R^n_+ \to R^n$ is a continuously differentiable function such that the Jacobian associated with F is a P-matrix for every $x \in R^n_+$ then the CP associated with F+ q has at most one solution for every $q \in R^n$.

(b) If $F:R^n_+ \to R^n$ is a differentiable mapping such that all the principal minors of the Jacobian matrix of F are bounded between δ and δ^{-1} for some $0 < \delta < 1$, then the CP associated with F is globally uniquely solvable.

Proof (a) : From the fundamental theorem of Gale and Nikaido it follows that F is a P-function in R^n_+. That is for every $x \neq y$ we have

$$\max(x_i - y_i)(f_i(x) - f_i(y)) > 0.$$

Consequently there can be at most one solution to the complimentarity problem,

$$x \geq 0,\ F(x) \geq 0 \quad \text{and} \quad x.F(x) = 0$$

This can be seen as follows. Suppose $x \neq y$ with

$$x \geq 0,\ F(x) \geq 0 \quad \text{and} \quad x.F(x) = 0$$

$$y \geq 0,\ F(y) \geq 0 \quad \text{and} \quad y.F(y) = 0 .$$

Consider for any i, $(x_i - y_i)(f_i(x) - f_i(y))$

$$= x_i f_i(x) + y_i f_i(y) - x_i f_i(y) - y_i f_i(x)$$

$$= - x_i f_i(y) - y_i f_i(x) \leq 0$$

This contradicts the fact that F is a P-function. Since F is a P-function F+ q is also a P-function for any $q \in R^n$. Consequently CP associated with F+ q also has at most one solution. This terminates the proof of (a) of theorem 2.

Proof (b) : We will now prove that the extension map G of F is a homeomorphism. Then from theorem 1 it will follow that F is globally uniquely solvable. Recall $G(z) = F(z^+) + z^-$ for any $z \in R^n$. Observe that the mappings $x \to x^+$ and $x \to x^-$ are differentiable in points x such that $x_i \neq 0$ $(i = 1,2,\ldots,n)$, it follows that G is differentiable in the interior of every orthant of R^n. However we can approximate G by differentiable functions $G_\alpha(x)$ and then apply Gale-Nikaido's theorem to these functions. Specially, for any $\alpha > 0$ and any real number ξ let

$$t(\xi,\alpha) = \begin{cases} 0 & \text{if} \quad \xi \leq -\alpha \\ (\xi + \alpha)^2/4\alpha & \text{if} \quad -\alpha < \xi \leq \alpha \\ \xi & \text{if} \quad \alpha < \xi \end{cases}$$

Denote by $u_i(x,\alpha) = t(x_i,\alpha)$, $i = 1,2,\ldots,n$, and $u = (u_1,u_2,\ldots,u_n)$ and define $G_\alpha(x) = F(u(x,\alpha))-u(-x,\alpha)$. One can check that the principal minors of the Jacobian matrices of the functions $G_\alpha(x)$ lie between δ and δ^{-1} $(\delta > 0)$. First we will show that G is onto. Note that from theorem 6 of chapter VI each G_α is onto. That is given any $q \in R^n$ we can find for each α, an x^α such that $G_\alpha(x^\alpha) = q$. If $\alpha \in (0,1]$, $\{x^\alpha\}$ is bounded. Hence it follows that for some $x, G(x) = q$. Now we will show that G is one-one. From Gale-Nikaido's theorem each G_α is one-one. Consequently G is one-one in the interior of every orthant in R^n. We will now establish that G is globally one-one in R^n. To achieve this end, for any $x \in R^n$, denote $I(x)=\{i:x_i=0\}$. Let $x,y \in R^n$ be such that $G(x) = G(y)$. We shall prove by induction on $|I(x)|+|I(y)|$ that $x = y$. Suppose first that $|I(x)| + |I(y)| = 0$ (Here $|I(x)|$ stands for the cardinality of the set $I(x)$). This means x and y are interior points of some orthants; that is, $x \in \text{int } Q^S$, $y \in \text{int } Q^T$ where S and T are subsets of $N = \{1,2,\ldots,n\}$ with $Q^S = \{x:x \in E^n,\ x_i \geq 0\ \forall\ i \in S \text{ and } x_i \leq 0 \text{ for all } i \in N\setminus S\}$etc. Thus we can find $u \in \text{Int } Q^S$ and $v \in \text{int } Q^T$ such that $G_\alpha(u) = G(x)$ and $G_\alpha(v) = G(y)$ for some $\alpha > 0$. Since G_α is univalent in R^n, $u = v$ and consequently $S = T$. Hence it follows that $x = y$ since G is univalent in the interior of every orthant.

Assume, by induction, that $G(x) = G(y)$ and $|I(x)| + |I(y)| \leq k$ imply $x = y$. Suppose $x \neq y$ with $G(x) = G(y)$ and $|I(x)| + |I(y)| = k+1$. Without loss of generality, suppose $|I(x)| \leq |I(y)|$. We shall show first G is locally univalent. Let N_x denote a neighbourhood of x such that $y \notin N_x$ and $I(u) \subset I(x)$ for every $u \in N_x$. We claim that G is univalent in N_x. Let u and v be two distinct points of N_x. We distinguish two cases. First $I(u) = I(x)$. In this case there exists an orthant Q^S such that $x,u,v \in Q^S$ and since G is univalent in every orthant, $G(u) \neq G(v)$. Second, $I(u) \subsetneqq I(x)$. In this case $|I(u)| + |I(v)| \leq 2|I(x)| \leq k+1$, so that the induction hypothesis implies $G(u) \neq G(v)$. Hence G is univalent on N_x.

It follows from the invariance theorem of domain that $G(N_x)$ is an open set which contains $G(y)$. Since G is continuous at y, there exists $w \notin N_x$ such that $I(w) = \phi$ and $G(w) \in G(N_x)$. Thus, we can find a $u \in N_x$ such that $G(w) = G(u)$. Since $|I(w)| + |I(u)| < k+1$, it follows from the induction hypothesis that $w = u$ which is a contradiction. This shows that G is one-one throughout R^n. Hence G is a homeomorphism of R^n onto R^n. Consequently from theorem 1, F is globally uniquely solvable. This completes the proof of part (b) of theorem 2.

<u>Remark</u> : Proof of part (b) of theorem is due to Megiddo and Kojima. As a corollary one can prove the following result due to Samuelson, Thrall and Wesler.

<u>Corollary</u> : Let $F(z) = Az + b$ be an affine mapping from R^n into itself. Then the CP associated with F is globally uniquely solvable if and only if all the principal minors of A are positive.

Proof : "If" part follows from theorem 2. We will only prove the "only if" part. That is we will assume F is globally uniquely solvable and show that A is a P-matrix. We will show that A does not reverse the sign of any nontrivial vector in R^n. Suppose there exists an $x \in R^n$, $x \neq 0$ such that $x_i(Ax)_i \leq 0$ for $i = 1,2,\ldots,n$. Let us write $y_i = (Ax)_i$ - ith component of the vector Ax. Write $y_i^+ = y_i$ if $y_i > 0$ and $y_i^+ = 0$ if $y_i \leq 0$; $y_i^- = 0$ if $y_i \geq 0$ and $y_i^- = -y_i$ if $y_i \leq 0$. Similarly we write x_i^+ and x_i^-. Then,

$$y_i = y_i^+ - y_i^- \,, \quad y_i^+, y_i^- \geq 0 \quad \text{and} \quad y_i^+ y_i^- = 0 \quad \text{for } i = 1,2,\ldots,n$$

$$x_i = x_i^+ - x_i^- \,, \quad x_i^+, x_i^- \geq 0 \quad \text{and} \quad x_i^+ x_i^- = 0 \quad \text{for } i = 1,2,\ldots,n \,.$$

Since $x_i y_i \leq 0$ for each $i = 1,2,\ldots,n$, it follows that $x_i^+ y_i^+ = 0 = x_i^- y_i^-$ for $i = 1,2,\ldots,n$. Hence $y^+ \cdot x^+ = y^- \cdot x^- = 0$ [Here dot refers to the inner product between the vectors]. Since $y = Ax$, we have $y^+ - Ax^+ = y^- - Ax^- = q_o$ (say). Since $x \neq 0$, $x^+ \neq x^-$. Hence one can conclude from,

$$Ax^+ + q_o = y^+ \quad \text{and} \quad x^+ \cdot y^+ = 0$$

$$Ax^- + q_o = y^- \quad \text{and} \quad x^- \cdot y^- = 0 \,.$$

$Ax + q_o$ has two solutions which contradicts our assumption that CP associated with $Ax + q_o$ has a unique solution. Hence A is a P-matrix. This terminates the proof of the corollary.

In theorem 2(b) we have shown that the CP-associated with F is globally solvable provided all the principal minors of the Jacobian matrix of F are bounded between δ and δ^{-1} for some $0 < \delta < 1$. The following example demonstrates that this result may fail without this condition. Let $F = (f,g,h)$ be a map from R_+^3 to R^3 where

$$f(x_1,x_2,x_3) = (e^{x_1} - e^{x_2} - e^{-x_2} + 1)e^{x_3}$$

$$g(x_1,x_2,x_3) = (e^{x_2} - e^{x_1} - e^{-x_1} + 1)e^{x_3}$$

$$h(x_1,x_2,x_3) = x_3 \,.$$

In this case Jacobian matrix turns out to be

$$J = \begin{bmatrix} e^{x_1} e^{x_3} & (-e^{x_2} + e^{-x_2})e^{x_3} & e^{x_3}(e^{x_1} - e^{-x_2} - e^{x_2} + 1) \\ (-e^{x_1} + e^{-x_1})e^{x_3} & e^{x_3} e^{x_2} & e^{x_3}(e^{x_2} - e^{x_1} - e^{-x_1} + 1) \\ 0 & 0 & 1 \end{bmatrix}$$

Clearly all the principal minors of J are all bounded below by one but some principal minors (for example $e^{x_1} e^{x_3}$) are not bounded above. Thus the conditions imposed on theorem 2(b) are not met. We will now show that the CP associated with F is not globally uniquely solvable. Let $q = (e,e,1)$ and $G = F(x_1,x_2,x_3)-q$. Then the CP associated with G is not solvable: Suppose the CP associated with G has a solution. This means that that there exists a $z \in R^3_+$ such that $G(z) \in R^3_+$ with $z.G(z) = 0$. Let $z = (z_1,z_2,z_3)$. Since $G(z) \in R^3_+$, it follows that $z_3-1 \geq 0$ or $z_3 \geq 1$. Since $z.G(z) = 0$, and $z_3 \geq 1$ third coordinate of $G(z) = 0$ or $z_3-1 = 0$ or $z_3 = 1$. Now we have,

$$(e^{z_1} - e^{z_2} - e^{-z_2} + 1)e \geq e$$

and

$$(1 + e^{z_2} - e^{z_1} - e^{-z_1})e \geq e$$

since $G(z) \in R^3_+$. From these inequalities we have,

$$e^{z_1} - e^{z_2} - e^{-z_2} \geq 0$$

and

$$e^{z_2} - e^{z_1} - e^{-z_1} \geq 0 .$$

Adding these two inequalities we get, $-(e^{-z_1} + e^{-z_2}) \geq 0$ which is impossible and thus we arrive at a contradiction to the supposition that the CP associated with G has a solution. Thus in this example F is not globally uniquely solvable.

Note that the CP associated with $F+q$ for any $q \in R^3$ is feasible, that is, there exists $z \geq 0$ such that $F(z) + q \geq 0$. This is met here since $\lim_{a \to \infty} F(1,1,a) = (\infty,\infty,\infty)$.

<u>An application of Hadamard's inverse function theorem to Algebra [22]</u> : By using a result of Hadamard we will demonstrate that Euclidean n-space cannot be endowed with the structure of a commutative division algebra when $n \geq 3$. More precisely we consider the possibility of defining an operation of "multiplication" $(x,y) \to xy$ of R^n which obeys the following axioms :

(i) $x(\lambda y) = \lambda xy$ where λ is a scalar.

(ii) $x(y+z) = xy + xz$

(iii) $xy = 0 \implies x = 0$ or $y = 0$

(iv) $xy = yx$.

We now have the following theorem.

<u>Theorem 3</u> : For $n \geq 3$ there is no operation of multiplication on R^n which satisfies (i) - (iv).

Remark : It is clear that when n = 1 or 2, we can define a multiplication operation satisfying (i) through (iv) and also associative law. Theorem 3 says that it is not possible to do so in R^n for $n \geq 3$. We will use a special case of a theorem due to Hadamard to prove theorem 3 - this proof is due to Gordon [21, 22].

Hadamard's theorem : Let $F:R^n - \{0\} \to R^n - \{0\}$ be a $C^{(1)}$ differentiable map with $n \geq 3$. Then F is a diffeomorphism of $R^n - \{0\}$ onto $R^n - \{0\}$ if and only if F is proper and the Jacobian of F never vanishes. [For a proof of Hadamard's theorem see [21]].

The following two examples show that Hadamard's theorem may fail to hold good when n = 1 or n = 2. When n = 1, $f(x) = x^2$ is a proper map with derivative non-zero for $x \neq 0$ and clearly it is not one-one. When n = 2, define $F(x_1,x_2) = (f,g)$ where $f(x_1,x_2) = x_1^2 - x_2^2$ and $g(x_1,x_2) = 2x_1x_2$. One can easily check that F is proper and its Jacobian is non-vanishing in $R^2-\{0\}$. Also F is not one-one since $F(1,1) = F(-1,-1)$.

Remark : Proof of Hadamard's theorem depends on the fact that $R^n-\{0\}$ is simply connected when $n \geq 3$.

Proof of theorem 3 : (Proof due to Gordon) We will show that axioms (i) through (iv) imply that the map $x \to x^2$ is a homeomorphism when $n \geq 3$ (which is obviously absurd since axiom (i) requires that $(-x)^2 = x^2$).

Let F be the map from $R^n - \{0\}$ to $R^n - \{0\}$ where $F(x) = x^2$. This is a well-defined map because of axiom (iii). Using the relevant axioms one can easily check that F is continuous and proper. Using axiom (iv) we can compute $dF_x(v)$ (= the differential F and x operating on v) and it is seen that

$$dF_x(v) = \lim_{h \to 0} \{ \frac{1}{h} (F(x+hv)-F(x))\} = xv + vx = 2xv.$$

Hence axiom (iii) implies that dF_x is non-singular for all $x \in R^n-\{0\}$ so that $F(x) = x^2$ is a homeomorphism via Hadamard's theorem which of course leads to a contradiction. Hence there cannot exist a multiplication operation in R^n for $n \geq 3$ satisfying the four axioms. This terminates the proof of theorem 3.

On the infinite divisibility of multivariate gamma distributions : In this section we will consider the problem of deciding whether a certain multivariate gamma distribution is infinitely divisible. We will give a sufficient condition due to Paranjape. In this situation weak N-matrices play an important role.

Let $X = (X_1,X_2,\ldots,x_p)$ be a multivariate normal random vector with zero mean vector and positive definite variance covariance matrix Σ. The characteristic function of $\left(\frac{x_1^2}{2}, \frac{x_2^2}{2}, \ldots, \frac{x_p^2}{2}\right)$ is given by $h_p(t) = h(t_1,t_2,\ldots,t_p) = |I-T\Sigma|^{-\frac{1}{2}}$ where T is a diagonal matrix with diagonal elements $it_1,\ldots,it_p$, $i = \sqrt{-1}$. Call $h_p(t)$ infinitely divisible if $(h_p(t))^a$ is a characteristic function for every $a > 0$. Paul Levy conjectured that $h_p(t)$ is not infinitely divisible. However Vere-Jones proved $h_2(t)$ is infinitely divisible, and Moran and Vere-Jones established $h_p(t)$ is infinitely divisible if $\Sigma = (1-\rho)I + \rho E_{pp}$ with $\rho > 0$, I = identity matrix and E_{pp} matrix with every entry equal to one. Also Griffiths obtained a necessary and sufficient conditions for $h_3(t)$ to be infinitely divisible. We will present a result due to Paranjape which in some sense unifies the known results. The problem posed by Levy is still open. We will now prove the following:

<u>Infinite divisibility theorem</u> : If for a set of positive constants $c_1,c_2,\ldots,c_p$, the matrix $(\text{diag}\,(c_1,c_2,\ldots,c_p)\Sigma^{-1} - I)$ is a weak N-matrix (that is all the principal minors are non-positive) then $h_p(t)$ is infinitely divisible.

<u>Remark</u> : If f(t) is a characteristic function then for $0 < \mu < 1$, $(1-\mu)/(1-\mu(f(t))$ is infinitely divisible. If $c_1,c_2,\ldots,c_p$ are positive constants then $h_p^2(t)$ can be written as

$$h_p^2(t) = \prod_{j=1}^{p} g_j(t_j)|\Sigma^{-1}| \prod_{i=1}^{p} c_i/\det\,(I+ A)$$

where $g_j(t_j) = (1-ic_jt_j)^{-1}$ and $A = (\text{diag}(c_1,c_2,\ldots,c_p)\Sigma^{-1}-I)\text{diag}(g_1(t_1),g_2(t_2),\ldots, g_p(t_p))$.

<u>Proof</u> : (of infinite divisibility theorem). Observe that $\det\,(I+ A) = I + \sum_{i=1}^{p} \text{tr}_i A$ where $\text{tr}_i A$ denotes the sum of all principal minors of order i, $1 \leq i \leq p$ of the matrix A. Write $P = \text{diag}(c_1,c_2,\ldots,c_p)\Sigma^{-1} - I$. A typical rth order principal minor of A is equal to

$$\prod_{j=1}^{r} g_{k_j}(t_{k_j})\, P(k_1,k_2,\ldots,k_r)$$

where $P(k_1,k_2,\ldots,k_r)$ is the $(k_1,k_2,\ldots,k_r)$-th principal minor of P. Set $\lambda = \sum_{r=1}^{p} \sum_{(k_1,\ldots,k_r)} (-P(k_1,k_2,\ldots,k_r))$. Since each $P(k_1,k_2,\ldots,k_r) \leq 0$ by hypothesis of the theorem, it follows that $\lambda > 0$. Therefore

$$\det(I+A) = 1-\lambda \left[\sum_{r=1}^{p} \sum_{(k_1,\ldots,k_r)} (-P(k_1,\ldots,k_r)/\lambda) \prod_{j=1}^{r} g_{k_j}(t_{k_j}) \right]$$

Since $-P(k_1,\ldots,k_r)/\lambda \geq 0$ and $\Sigma(-P(k_1,k_2,\ldots,k_r))/\lambda = 1$, it follows that the expression in the right hand side which appears within the brackets in the expansion of $\det(I+A)$, is a multivariate characteristic function. Define

$$h_p^*(t) = \det(\Sigma^{-1}) \prod_{i=1}^{p} c_i / \det(I+A).$$

Since $h_p(0) = 1$, $h_p^*(0) = 1$ and consequently $\det(\Sigma^{-1}) \prod_{i=1}^{p} c_i/(1-\lambda) = 1$. Since the numerator is positive, $1-\lambda > 0$ or $\lambda < 1$. Thus $h_p^*(t)$ has the representation of the form $(1-\lambda)/(1-\lambda f(t))$ where $f(t)$ is a characteristic function and therefore $h_p^*(t)$ is infinitely divisible. Consequently $h_p^2(t)$ is infinitely divisible or $h_p(t)$ is infinitely divisible. This terminates the proof of the theorem on infinite divisibility of $h_p(t)$.

Remark : The following result due to Rao gives an equivalent condition for a matrix A to be a weakly N-matrix. A nonpositive symmetric matrix $A(\neq 0)$ is merely positive subdefinite if and only if it is a weakly N-matrix. For a proof see Rao [62] [Call a real symmetric matrix A positive subdefinite if for any vector x, $x'Ax < 0$ implies $Ax \leq 0$ or $Ax \geq 0$. A positive submatrix which is not positive semidefinite is called a merely positive subdefinite matrix].

Examples : We will prove $h_2(t)$ is always infinitely divisible by actually checking the condition given in the theorem. If $\Sigma^{-1} = \begin{bmatrix} a & b \\ b & c \end{bmatrix}$ choose $c_1 = \frac{1}{a}$, $c_2 = \frac{1}{c}$ one can easily check, $\text{diag}\,(c_1,c_2)\Sigma^{-1}-I = \begin{bmatrix} 0 & b/a \\ b/c & 0 \end{bmatrix}$. Clearly principal minors are non-positive. Hence $h_2(t)$ is infinitely divisible. As another example we will look at Griffith's result in three dimensions where

$$\Sigma = (1-\rho)I + \rho E_{33} = \begin{bmatrix} 1 & \rho & \rho \\ \rho & 1 & \rho \\ \rho & \rho & 1 \end{bmatrix}$$

$$\Sigma^{-1} = \begin{bmatrix} \frac{1-\rho^2}{(1-\rho)^2(1+2\rho)} & \frac{\rho^2-\rho}{(1-\rho)^2(1+2\rho)} & \frac{\rho^2-\rho}{(1-\rho)^2(1+2\rho)} \\ \frac{\rho^2-\rho}{(1-\rho)^2(1+2\rho)} & \frac{1-\rho^2}{(1-\rho)^2(1+2\rho)} & \frac{\rho^2-\rho}{(1-\rho)^2(1+2\rho)} \\ \frac{\rho^2-\rho}{(1-\rho)^2(1+2\rho)} & \frac{\rho^2-\rho}{(1-\rho)^2(1+2\rho)} & \frac{1-\rho^2}{(1-\rho)^2(1+2\rho)} \end{bmatrix}$$

Choose $c_1 = c_2 = c_3 = (1-\rho)^2 (1+2\rho)/(1-\rho^2)$. Then $(\mathrm{diag}(c_1,c_2,c_3)\Sigma^{-1}-I)$ will be a weak N-matrix provided $\rho > 0$. Therefore $h_3(t)$ will be infinitely divisible provided Σ is of the form $(1-\rho)I + \rho E_{33}$ where $\rho \geq 0$. In general it is not known whether $h_3(t)$ is infinitely divisible.

CHAPTER X

FURTHER GENERALIZATIONS AND REMARKS

Abstract : In this chapter we will first discuss a generalization of local inverse function theorem due to Clarke and Hadamard's theorem due to Pourciau when F is a Lipschitzian function but not necessarily a $C^{(1)}$ function. We will then discuss the notion of monotone functions. Next we will say something about PL functions. Finally we will discuss global univalent results when the Jacobian is allowed to vanish. Main aim of this chapter is to indicate possible generalizations in global univalent results.

A generalization of the local inverse function theorem : Here we will describe a nice generalization of the local inverse function theorem due to Clarke. Let $F : R^n \to R^n$ satisfy a Lipschitz condition in a neighbourhood of a point y_0 in R^n. That is for some constant d, for all x and y near y_0, we have

$$||F(x)-F(y)|| \leq d||x-y||$$

where $||\cdot||$ denotes as usual the Euclidean norm. Let J denote the Jacobian whenever the partial derivatives exist. We will metrize the vector space M of n x n matrices with the norm,

$$||A|| = \max |a_{ij}|$$

where

$$A = (a_{ij}),\ 1 \leq i \leq n \quad \text{and} \quad 1 \leq j \leq n.$$

Definition : The generalized Jacobian of F at y_0 denoted by $\partial F(y_0)$, is the convex hull of all matrices A of the form,

$$A = \lim_{n \to \infty} J(y_n)$$

where y_n converges to y_0 and F is differentiable at y_n for each n.

Remark 1 : It follows from Rademacher's theorem that F is almost everywhere differentiable near y_0. Furthermore J(y) is bounded near y_0 since F satisfies a Lipschitz condition in a neighbourhood of y_0.

Remark 2 : The generalized Jacobian $\partial F(y_0)$ is a nonempty compact convex subset, in the vector space M of matrices.

Remark 3 : If F is $C^{(1)}$, $\partial F(y_0)$ reduces to $J(y_0)$.

Definition : The generalized Jacobian $\partial F(y_0)$ is said to be of maximal rank if every A in $\partial F(y_0)$ is of maximal rank.

We are ready to state the theorem due to Clarke.

Generalized local inverse function theorem : If $\partial F(y_0)$ is of maximal rank, then there exist open sets U and V of y_0 and $F(y_0)$ respectively, and a Lipschitzian function $G:V \to R^n$

(i) $G(F(u)) = u$ for every $u \in U$

(ii) $F(G(v)) = v$ for every $v \in V$

For a proof of this result see Clarke [12]. In view of remark 3, it is clear that when F is $C^{(1)}$, G is necessarily $C^{(1)}$. Also observe that it is not sufficient to assume that J is of maximal rank whenever it exists, as the function $|x|$, (n=1) demonstrates. A simple example to which the above theorem applies, n = 2, is the following: $F(x,y) = (|x|+y,\ 2x+|y|)$ near (0,0). Here one can check that,

$$\partial F(0,0) = [\ [\begin{smallmatrix} s & 1 \\ 2 & t \end{smallmatrix}] : -1 \leq s \leq 1,\ -1 \leq t \leq 1]$$

Immediately one can raise the following question in view of this result due to Clarke : Is it possible to formulate a global univalent result when F satisfies Lipschitz condition but F not necessarily a $C^{(1)}$ function? In fact Pourciau recently has shown that Hadamard's theorem holds good for locally Lipschitzian maps. In order to state the theorem due to Pourciau [58] we need to slightly extend the concept of generalized Jacobian that was given earlier.

Let $F:R^n \to R^n$. Call F locally Lipschitzian provided each point x has a neighbourhood U where some positive number M satisfies $||F(z_1)-F(z_2)|| \leq M||z_1-z_2||$ for all $z_1, z_2 \in U$. As already remarked $F'(x)$ exists a.e. with respect to Lebesgue measure. Moreover almost every x is a Lebesgue point of the derived mapping F'. By definition such x satisfy

$$\lim_{\varepsilon \to 0} \frac{1}{\ell(B_\varepsilon(x))} \int_{B_\varepsilon(x)} ||F'(z)-F'(x)||\,d\mu(z) = 0$$

Here $\ell(B_\varepsilon(x))$ stands for the Lebesgue measure of $B_\varepsilon(x)$ (= Ball of radius ε with centre x). Let $L(F')$ stand for the set of all these Lebesgue points and let $p \in R^n$. Then the generalized derivative $\partial F(p)$ of F at p is the non-empty, compact convex subset $\bigcap_{\delta>0} \Delta_\delta(F,p)$ of $L(R^n,R^n)$, where $\Delta_\delta(F,p)$ denotes the collection

$$\overline{\text{Con}}\ \{F'(x) : x \in B_\delta(p) \cap L(F')\} .$$

This extra condition allows us to ignore null sets in forming the generalized derivative. With this notion of generalized derivative, Pourciau has proved the local inverse function theorem stated in chapter I for Lipschitz functions. For details see [57]. We are now ready to state the following:

Lipschitzian Hadamard Theorem : Suppose F is a locally Lipschitzian map from R^n into R^n and let $M > 0$. If $\partial F(p)$ is invertible and $||A^{-1}|| \leq M$ for each $p \in R^n$ and each A in $\partial F(p)$, then F is a homeomorphism from R^n onto R^n.

For a proof of this result see [58].

Remark 1 : It appears that some of the results proved in chapter IV, in particular Plastock's results might be suitably formulated for locally Lipschitzian maps.

Remark 2 : Hadamard's theorem is generalized to Banach spaces by Caccioppoli [6] and Plastock has proved some of his results for Banach spaces. As such can we assert that Lipschitzian Hadamard's theorem holds good for Banach spaces?

Remark 3 : Is it possible to formulate Mas-Colell's or Garcia-Zangwill's global univalent result for locally Lipschitzian maps ?

Monotone functions and univalent mappings : In the introduction we have mentioned as one of the approaches for tackling global univalent problem, is via monotone functions. In this section we will present an example of that approach.

Definition : A mapping $F:R^n \to R^n$ is monotone if $(F(x)-F(y))'(x-y) \geq 0$ for every $x,y \in R^n$. Call F strictly monotone if $(F(x)-F(y))'(x-y) > 0$ whenever $x \neq y$. Call F uniformly monotone if there is a $\delta > 0$ so that $(F(x)-F(y))'(x-y) \geq \delta(x-y)(x-y)$ for all $x,y \in R^n$.

[Here as usual prime denotes transpose].

This concept is a natural non-linear generalization of positive definiteness. The following proposition is easy to prove.

Proposition 1 : Let Ω be an open convex set in R^n. Let $F:\Omega \to R^n$ be a $C^{(1)}$ map. Then

(i) F is monotone if and only if J is positive semi-definite for all $x \in \Omega$.

(ii) If J is quasi-positive definite for all $x \in \Omega$, then F is strictly monotone on Ω.

(iii) F is uniformly monotone on Ω if and only if there is a $\delta > 0$ so that $h'Jh \geq \delta h'h$ for all $x \in \Omega$ and $h \in R^n$.

Proof : We will prove (i) and (iii) simultaneously. Suppose F is uniformly monotone. Then for any $x \in \Omega$ and $h \in R^n$, we have

$$h'J(x)h = h'\lim_{t\to 0} [F(x+th)-F(x)]$$

$$\geq \lim_{t\to 0} t^{-2}\,\delta\|th\|^2 = \delta h'h \quad .$$

If F is monotone then $\delta = 0$ and $J(x)$ is positive semi-definite. Conversely suppose $h'J(x)h \geq \delta h'h$. Then from the mean-value theorem we have,

$$(x-y)'F(x)-F(y)) = \int_0^1 (x-y)'J(y+t(x-y))(x-y)dt$$

$$\geq \delta(x-y)'(x-y)$$

so that F is monotone or uniformly monotone depending on whether $\delta = 0$ or $\delta > 0$. Finally, if $J(x)$ is quasi-positive definite for all $x \in \Omega$, and $x \neq y$, then the above integrand is positive for all $t \in [0,1]$ and hence F is strictly monotone. This terminates the proof of Proposition 1.

Proof of this proposition is taken from Ortega and Rheinboldt [48].

Remark : Observe that (ii) of Proposition 1 is given in Chapter III. However the conditions do not suffice for the existence of solutions as the example $F(x) = e^x$ in one-dimension shows. We may guarantee existence, however, by strengthening the monotonicity assumption.

Theorem 1 : If $F:R^n \to R^n$ is a $C^{(1)}$ map and uniformly monotone on R^n then F is one-one and onto R^n and consequently F is a homeomorphism.

From the given hypothesis one can easily verify that $\|J(x)^{-1}\| \leq \frac{1}{\delta}$ for all $x \in R^n$. [Here the norm of the matrix is taken as the ℓ_2 norm]. Now Hadamard's theorem will imply that F is one-one and onto. For details of the proof see [42 or 48].

Remark 1 : If F is continuous and montone on the open, convex set Ω, then for any $b \in R^n$, the solution set $S = \{x \in \Omega : F(x) = b\}$ is convex if it is not empty.

Remark 2 : Suppose that $F, G : R^n \to R^n$ are both $C^{(1)}$ maps and montone and that F is uniformly monotone. Then the map H defined by : $H = F + G$ is a homeomorphism of R^n onto R^n. In particular if A is a quasi-positive definite matrix, $H = A + G$ is a homeomorphism. (See problem E 5.4-5 in [48]).

On PL-functions : Because of the growing importance (as well as simplicity) of piecewise linear functions, in this section we will prove a result due to Kojima-Saigal [32]. In order to do that we need some preliminaries.

Let S be a closed convex polyhedral subset of R^n, and let Σ be a class of closed convex polyhedral subsets of S which partition S. We will assume Σ contains only finitely many members.

Definition : Call (S,Σ) a subdivided polyhedron of dimension n if :

(a) elements of Σ are n-dimensional convex and polyhedral and are called pieces.

(b) any two members of Σ are either disjoint or meet on a common face.

(c) the union of the pieces in Σ is S.

For our result $S = R^n$. Let (R^n,Σ) be a subdivided polyhedron, and let

$$F : R^n \to R^n$$

be piecewise linear and continuously differentiable on this subdivision, that is, PC^1 with affine on each piece of Σ. Since Σ contains a finite number of pieces, outside some compact region, points of R^n lie in some unbounded piece in Σ. Let these unbounded pieces be numbered $\sigma_1,\sigma_2,\ldots,\sigma_k$ and let $F(x)|_{\sigma_i} = A_i x - a_i$ for some n x n matrices A_i, and n vectors a_i. Then we have the following:

<u>Theorem 2</u> : Suppose that the Jacobian matrix of each piece of linearity of F has a positive determinant. Also, let there exist a matrix B such that $(1-t)B + tA_i$ is non-singular for each $t \in [0,1]$ and $i = 1,2,\ldots,k$. Then F is a homeomorphism of R^n onto R^n.

<u>Remark</u> : Let $z \in R^n$, such that $\det J_{F_\sigma}(z)$ is positive (negative) for every σ containing z. Then, there exists $\varepsilon > 0$ such that $\deg(F,B_\delta(z), F(z)) \geq 1$ (≤ -1) for each $\delta \in (0,\varepsilon)$. For a proof of this remark see [32].

<u>Proof of theorem 2</u> : Let $y \in R^n$ be arbitrary. Now consider the homotopy

$$H(x,t) = (1-t)Bx + t(F(x)-y),\ t \in [0,1].$$

We will first prove that $H^{-1}(0)$ has no unbounded component. Assume the contrary this means that for some σ_i we can find a sequence $(x^m,t_m) \in H^{-1}(0)$, $m = 1,2,\ldots$ such that $x^m \in \sigma_i$ and $||x^m|| \to \infty$. Also on some subsequence $x^m/||x^m|| \to x^*$, $t_m \to t^*$, $t^* \in [0,1]$ and $x^* \neq 0$. Hence we have

$$(1-t^*)Bx^* + t^* A_i x^* = 0$$

which is a contradiction to our hypothesis that $(1-t)B + tA_i$ is non-singular for each $t \in [0,1]$. Since $H^{-1}(0)$ is bounded for each y, and det $B > 0$, from the homotopy invariance theorem the degree of $F(x)-y$ is $+1$ for all y. From the above remark it follows that F is one-one and onto. This terminates the proof of theorem 2.

Now we will give two examples. In the first example conditions of theorem 2 are satisfied. In the second example one of the conditions of theorem 2 will be violated but the map will be a homeomorphism.

Example 1 : Consider $F:R^2 \to R^2$ where

$$F(x,y) = \begin{cases} (x-y,y) & \text{if} \quad x \geq 0 \\ (2x-y,y) & \text{if} \quad x < 0 \end{cases}$$

In this example $A_1 = \begin{bmatrix} 1 & -1 \\ 0 & 1 \end{bmatrix}$, $A_2 = \begin{bmatrix} 2 & -1 \\ 0 & 1 \end{bmatrix}$ and let $B = \begin{bmatrix} \frac{1}{2} & 0 \\ 0 & \frac{1}{2} \end{bmatrix}$. One can easily check that $(1-t)B + tA_i$ is non-singular for each $t \in [0,1]$ and $i = 1,2$. Hence F is a homeomorphism of R^2 onto R^2.

We need the following lemma for example 2.

Lemma : Let $A_1 = \begin{bmatrix} -1 & 0 \\ -2 & 1 \end{bmatrix}$, $A_2 = \begin{bmatrix} -1 & 0 \\ 2 & 1 \end{bmatrix}$, $A_3 = \begin{bmatrix} 1 & -2 \\ 0 & -1 \end{bmatrix}$ and $A_4 = \begin{bmatrix} 1 & 2 \\ 0 & -1 \end{bmatrix}$. Then there does not exist any matrix B such that $tB + (1-t)A_i$ is non-singular for every $t \in [0,1]$ and $i = 1,2,3,4$.

Proof : One can possibly give a direct proof but our proof depends on theorem 2. Consider the following diagram, which represents a piecewise linear function F in R^2.

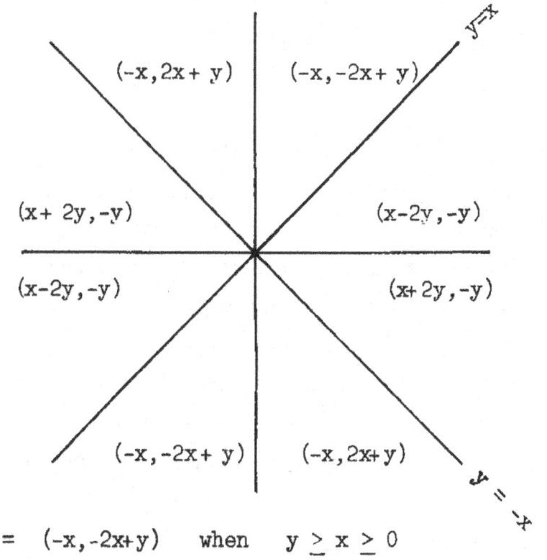

For example $F(x,y) = (-x,-2x+y)$ when $y \geq x \geq 0$

$= (x+2y, -y)$ when $0 \leq -y \leq x$.

Clearly the given A_1, A_2, A_3, A_4 are the Jacobians of F in different regions. Also note that $F(\frac{1}{2}, 1) = F(\frac{1}{2},-1) = (-\frac{1}{2}, 0)$ and consequently the given F is not a homeomorphism. Hence from theorem 2 we can conclude that there cannot exist a matrix B with $tB + (1-t)A_i$ nonsingular for every $t \in [0,1]$ and $i = 1,2,3,4$. This terminates the proof of the lemma.

We will use this lemma in our next example to show that the condition given in theorem 2 is not a necessary condition for an F to be a homeomorphism.

Example 2 : See the diagram below. In this example R^2 is divided into 6 regions and in each region we have given the Jacobian of the function F. Clearly F is a homeomorphism but this F does not satisfy the condition imposed in theorem 2. In other words this example demonstrates that the condition "$tB+ (1-t)A_i$ is non-singular for every $t \in [0,1]$ and $i = 1,2,\dots,k$ for some matrix B" is not a necessary condition for homeomorphism.

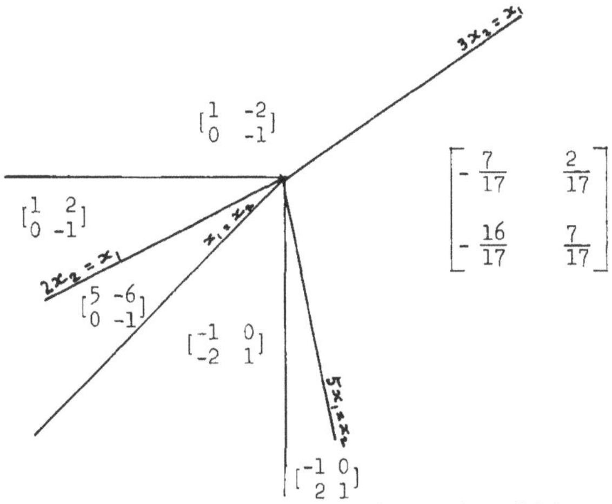

Because of the growing importance of PL-functions it would be nice to give univalence conditions that are both necessary and sufficient. Recently Schramm in fact has given in [70] necessary and sufficient conditions for PL-functions to be a homeomorphism. Among other results he proves that when the determinants associated with a PL-function $F:R^n \to R^n$ have the same sign, F is an open mapping. Also he is able to give bounds for the number of solutions of certain PL-equations. For details see [69]. Gale-Nikaido's results have also been extended by Kojima-Saigal for piecewise continuously differentiable, PC^1-functions [32].

On a global univalent result when the Jacobian vanishes : In this section we will give a global univalent result due to Chua and Lam [9] when the Jacobian is allowed to vanish on a set of isolated points. Kojima-Saigal have shown a similar result thereby extending Gale-Nikaido's fundamental theorem. For the proof of Chua and Lam we need the following results on local homeomorphism where the Jacobian may be allowed to vanish.

(i) Let F be C^1 map from an open set $U \subset R^n \to R^n$ and $n \geq 3$. Suppose $J(t) \neq 0$, $t \in U \setminus \{t_o\}$ where t_o is an isolated point. Then F is a local homeomorphism at t_o. This result is due to Church and Hemmingsen [11].

(ii) Let F be a $C^{(n)}$ map from R^n to R^n where $n \geq 3$. Let R_{n-1} denote the set of zeros of the Jacobian. If R_{n-1} is zero-dimensional then F is a local homeomorphism. This result is due to Church [10].

Remarks : It is interesting to note that both the results are valid when $n \geq 3$. For $n = 2$, the analytic function $f(z) = z^2$ is an obvious counter example. Proofs of (i) and (ii) depend on the fact that the set of branch points (= collection of points at which F fails to be a local homeomorphism) forms a perfect set when $n \geq 3$.

We are ready to state the following

Theorem 3 : (Chua and Lam) Let $F:R^n \to R^n$ be a C^1 map where $n \neq 2$. Let $I=\{x:x \in R^n$ and $J(x)$ is singular$\}$ and $I^c = \{x:x \notin I\}$. Then the following conditions are sufficient for F to be a homeomorphism of R^n onto R^n:

(1) $\det J(x) > 0$ for all $x \in I^c$ and I is at most a set of isolated points.

(2) F is norm-coercive.

Proof : First we will prove theorem 3 when $n = 1$. Suppose $F:R^1 \to R^1$ is not one-one This means we can find $x_1 \neq x_2$ with $F(x_1) = F(x_2)$. Condition (1) implies that there exists an interval (c,d) with $x_1 \leq c < d \leq x_2$ such that $F'(x) > 0$ on (c,d). Since F is C^1, we have

$$F(x_2)-F(x_1) = \int_{x_1}^{x_2} F'(x)dx \geq \int_c^d F'(x)dx > 0$$

which is a contradiction. Hence F is one-one and consequently F is a homeomorphism. Since F is norm coercive, F is onto R and this terminates the proof when $n = 1$.

We will now consider the case when $n \geq 3$. Let u be a point in I. There exists an open set N_u about u in R^n with $N_u \cap I = \{u\}$. Since $\det J(x) > 0$ on N_u except at the isolated point u where the Jacobian vanishes, it follows from the results quoted above that F is a local homeomorphism on N_u for each $u \in I$. Since $J(x)$ is non-singular for every $x \in I^c$, it follows that F is a local homeomorphism on R^n for all $n \geq 3$. Hence we can conclude that F is a homeomorphism of R^n (as F is norm-coercive and a local homeomorphism).

Proof of the theorem will be complete if we show that F is onto R^n. Suppose $F(R^n)$ is a proper subset of R^n. Observe $F(R^n)$ is open as R^n is open. Let $b \in R^n$ be a boundary point of $F(R^n)$ and let M_b an open connected neighbourhood of $F(b)$. Since F is a finite covering mapping on R^n, $F^{-1}(M_b)$ has a finite and a non-zero number of components. Let N_b be a component of $F^{-1}(M_b)$ that contains the point b. Let $N_b^* = F(R^n) \cap F^{-1}(M_b)$. Since F is continuous F^{-1} is open. Hence, both N_b and N_b^* are open and connected. Also observe that F maps both N_b and N_b^* topologically onto M_b. Clearly $N_b \cap F(R^n) \neq \emptyset$ as the point b belongs to both N_b and $F(R^n)$. It follows that $N_b \cap N_b^* \neq \emptyset$, for otherwise there will be at least one point x_1 in $N_b \cap F(R^n)$ and a point x_2 in N_b^* such that $F(x_1) = F(x_2) \in M_b$ and F restricted to $F(R^n)$ will not be one-one, which is a contradiction. As both N_b and N_b^* are connected, we have $N_b = N_b^*$. Hence it follows that $F(R^n)$ cannot be an open proper subset of R^n. That is $F(R^n)$, is closed in R^n. We have $F(R^n)$ which is both open and closed and it is non-empty. Therefore we can conclude that $F(R^n) = R^n$. This completes the proof of theorem 3.

Remark 1 : We have used theorem 3 in chapter V to prove theorem 2". We have given an example there which will serve as a counter example to theorem 3 in R^2. For extensions of Gale-Nikaido's univalent results and other results along these lines we refer the readers to [32], [9]. Another result which is closely related to theorem 3 is the following theorem due to Cronin and McAuley [14].

Theorem 4 : Let A and B be the interiors of two unit balls in R^n and F a continuously differentiable light and open map from $\bar{A}$ to $\bar{B}$. [Here $\bar{A}$ and $\bar{B}$ stand for the closure of A and B respectively]. Suppose $F(\bar{A} \setminus A) = \bar{B} \setminus B$ and $F|(\bar{A} \setminus A)$ is a homeomorphism. Let $J(x)$ stand for the Jacobian evaluated at $x \in A$ as usual, and let

$$J_o = \{x \in A : J(x) = 0\}$$

$$J_+ = \{x \in A : J(x) > 0\}$$

$$J_- = \{x \in A : J(x) < 0\}$$

$$J_1 = \bar{J}_+ \cap \bar{J}_- \cap A .$$

If $\dim J_1 < (n-1)$, then F is a homeomorphism.

Proof : See pp. 406-408 in Cronin and McAuley [14].

Remark 1 : This result enables us to study subsets of the set of points J_o where the Jacobian is zero instead of dealing directly with the branching or singular set which is a subset of J_o.

Remark 2 : The following example shows that there are many homeomorphic onto mappings with their Jacobian vanishing on an (n-1) dimensional set. The following simple example is a case in point.

Let $F:R^3 \to R^3$ be defined by

$$y_1 = f_1(x) = x_1^3$$

$$y_2 = f_2(x) = x_2^3$$

$$y_3 = f_3(x) = x_1 + x_2 + x_3.$$

This function F has a global inverse on R^3; namely

$$x_1 = y_1^{1/3} , \quad x_2 = y_2^{1/3}$$

$$x_3 = y_3 - y_1^{1/3} - y_2^{1/3}$$

Hence, $F : R^3 \to R^3$ is a homeomorphic onto mapping. But $\det J(x) = 9x_1^2\, x_2^2$ vanishes on two 2-dimensional hyperplanes : one defined by $x_1 = 0$ and the other defined by $x_2 = 0$. This shows that the condition "$\dim J_1 < n-1$" imposed in theorem 4 is not necessary.

Remark 3 : The set J_o contains the singular set or branch set - see McAuley [37]. But it $x \in J_o$ the map F can be locally one-to-one at x as the example of the map

$$y_1 = x_1^3$$

$$y_2 = x_2^3$$

in a neighbourhood of (0,0) shows.

Remark 4 : It may be possible to strengthen theorem 3 as well as theorem 4 with the help of theorem 6 (as well as theorem 7) of chapter IV but we have not made any attempt to do so.

Injectivity of quasi-isometric mappings : Let X,Y stand for real Banach spaces. Let Ω be an open subset of X. Following John, a mapping $F:\Omega \to Y$ is said to be (m,M)-isometric if it is a local homeomorphism (that is continuous, open and locally one-one) for which $M \geq D^+F(x)$ and $0 < m \leq D^-F(x)$ where $D^+F(x)$ and $D^-F(x)$ are respectively, the upper and lower limits $||F(y)-F(x)||/||y-x||$ as $y \to x$. Less precisely, F is called quasi-isometric if it is (m,M) - isometric for some m, M. We have then the following theorem due to Gevirtz.

Theorem 5 : Let X and Y be Banach spaces [In particular we can take $X = Y = R^n$]. Let $\Omega \subset X$ be an open ball and let $F:\Omega \to Y$ be an (m,M) isometric mapping. Then F is one-one if any one of the following conditions is satisfied.

(a) $M/m < \mu_o$ where μ_o is the unique real root of the equation $x = [x + (25x^2-8x)^{\frac{1}{2}}]/2(3x^2-x)$.

(b) X is a Hilbertspace and $M/m < \sqrt{2}$.

(c) X and Y are Hilbertspaces and $M/m < (1 + \sqrt{2})^{\frac{1}{2}}$.

For a proof of this result see J.Gevirtz [Proc. Amer. Math. Soc. 85 (1982), 345-349]. For related results see also F.John [On quasi-isometric mappings I, II Communications to Pure and Appl. Math. 1968, 1969 21,22 pp. 77-110, 265-278].

We will close this chapter by mentioning an old conjecture due to Jacobi. [This conjecture was brought to my attention by Professor Garcia]. Let $F:\mathbb{C}^n \to \mathbb{C}^n$ be a polynomial map. That is $F = (f_1,f_2,\ldots,f_n)$ where each f_i is a polynomial in n variables $z_1,z_2,\ldots,z_n$. If F has a polynomial inverse $G = (g_1,g_2,\ldots,g_n)$, then the determinant of the Jacobian matrix $\left(\frac{\partial f_i}{\partial z_j}\right)$ is a non-zero constant. This follows from the chain rule. Since F o G is the identity, we have $z_i = g_i(f_1,f_2,\ldots,f_n)$, so

$$\delta_{ij} = \frac{\partial}{\partial z_j} g_i(f_1,f_2,\ldots,f_n) = \sum_{t=1}^{n} \frac{\partial g_i}{\partial z_t}(f_1,\ldots,f_n)\cdot \frac{\partial f_t}{\partial z_j} .$$

This shows that the product

$$\left(\frac{\partial g_i}{\partial z_j}(f_1,\ldots,f_n)\right)\cdot\left(\frac{\partial f_i}{\partial z_j}\right)$$

is the identity matrix. Thus the Jacobian determinant of F is a non-vanishing polynomial, hence a constant. Now we are ready to state

The Jacobian conjecture : Let $F:\mathbb{C}^n \to \mathbb{C}^n$ be a polynomial map such that the Jacobian determinant is a non-zero constant. Then F has a polynomial inverse. For more information on this conjecture see the following articles:

D.Wright (1981), On the Jacobian conjecture, ILL.Jour. Math. 25, 423-439.

H.Bass, E.H. Connell and D. Wright (1982), The Jacobian conjecture - reduction of degree and formal expansion of the inverse, Bull. Amer. Math. Soc. 7, 287-330.

REFERENCES

[1] S.Banach and S.Mazur (1934), Uber mehradeutige stetige abbildungen, Studia Math 5, 174-178.

[2] M.S.Berger (1977) Non-linearity and functional analysis, Academic Press, New York.

[3] L.Bers (1956) Topology, Courant Institute of Mathematical Sciences, New York University, New York.

[4] W.E. Boyce and R.C. DiPrima (1977) Elementary differential equations, Third Edition, John Wiley and Sons, New York.

[5] F.E.Browder (1954) Covering spaces, fiber spaces and local homeomorphisms, Duke Math Jour 21, 329-336.

[6] R.Caccioppoli (1932) Sugli elementi uniti delle transformazioni funzionali:un teorema di esistenza e di unicita ed alucune sue applicazioni, Rend. Sem. Math. Padova 3, 1-15.

[7] P.Chander, The non-linear input-output model, To appear in Jour. Econ. Theory,

[8] A.Charnes, W.Raike and J.Stuz (1975) Poverses and the economic global unicity theorems of Gale and Nikaido, Zeitschrift for Operations Research 19, 115-121.

[9] L.O.Chua and Y.F.Lam (1972) Global homeomorphism of vector valued functions, Jour. Math. Analy and its Appl. 39, 600-624.

[10] P.T.Church (1963) Differentiable open maps on manifolds, Trans. Amer. Math. Soc. 109, 87-100.

[11] P.T.Church and E.Hemmingsen (1960) Light open maps on n-manifolds, Duke. Math. Jour. 27, 527-536.

[12] F.H.Clarke (1976) On the inverse function theorem, Pac. Jour. Math. 61, 97-102.

[13] J.Cronin (1964) Fixed points and topological degree in non-linear analysis, Providence Amer. Math. Soc.

[14] J.Cronin and L.F. McAuley (1966) Whyburn's conjecture for some differentiable maps, Proc. Natl. Acad. Sci.,U.S.A., 56, 405-412.

[15] Deimling (1974), Nichtlineare Gleichungen und Abbildungslehre, Springer-Verlag (Hochschul text).

[16] N.Dunford and J.T.Schwartz (1963) Linear Operators II, Spectral theory, Self-adjoint operators in Hilbert space, Interscience, New York.

[17] W.Fleming (1977) Functions of several variables, Second Edition, Springer Verlag, Heidelberg - New York.

[18] D.Gale (1960) The theory of linear ecomic models, Mcgraw Hill Book Company, New York.

[19] D.Gale and H.Nikaido (1965) The Jacobian matrix and global univalence of mappings, Math. Ann. 159, 81-93.

[20] C.B.Garcia and W.I.Zangwill (1979) On univalence and P-matrices, Lin. Alg. and its Appl. 24, 239-250.

[21] W.B.Gordon (1972) On the diffeomorphisms of euclidean space, Amer. Math. Monthly 79, 755-759.

[22] W.B.Gordon (1977) An application of Hadamard's inverse function theorem to Algebra, Amer. Math. Monthly 84, 28-29.

[23] J.Hadamard (1906) Sur les transformations ponctuelles, Bull. Soc. Math. France 34, 71-84.

[24] J.Hadamard (1910) Sur quelques applications de d'indice de Kronecker in "Introduction a la Theorie des Fonctions d'une Variable" by J.Tannery, pp. 437-477. Herman, Paris.

[25] P.Hartman (1961) On stability in the large for systems of ordinary differential equations, Can. J. Math. 13, 480-492.

[26] P.Hartman (1964) Ordinary differential equations, John Wiley, New York.

[27] P.Hartman and C.Olech (1962) On global asymptotic stability of solutions of differential equations, Trans. Amer. Math. Soc. 104, 154-178.

[28] R.Hermann (1968) Differential geometry and the calculus of variations, Math. in. Sci and Engineering Vol. 49, Academic Press, New York.

[29] R.V.Hogg and A.T.Craig (1970) Introduction to Mathematical Statistics, The Macmillan Company, Collier - Macmillan Limited, London pp 139-141.

[30] W.Hurewicz and H.Wallman (1948) Dimension Theory, Princeton University Press.

[31] H.Kestelman (1971) Mapping with non-vanishing Jacobian, Amer. Math. Monthly 78, 662-663.

[32] M.Kojima and R.Saigal (1979) A study of $PC^{(1)}$ homeomorphisms on subdivided polyhedrons, SIAM J. Math. Anal. 10, 1299-1312.

[33] C.Kosniowski (1980) A first course in algebraic topology, Cambridge University Press.

[34] L.Markus and H.Yamabe (1960) Global stability criteria for differential systems, Osaka Math. J. 12, 305-317.

[35] A.Mas-Colell (1979) Homeomorphisms of compact convex sets and the Jacobian matrix, SIAM J. Math. Anal. 10, 1105-1109.

[36] A.Mas-Colell (1979) Two propositions on the global univalence of systems of cost function, General Equilibrium, Growth and Trade, Edited by J.Green and J.Scheinlsman, Academic Press, New York.

[37] L.F.McAuley (1965) Concerning a conjecture of Whyburn on light open mappings, Bull. Amer. Math. Soc. 71, 671-674.

[38] L.F.McAuley (1966) Conditions under which light open mappings are homeomorphisms, Duke Math. Jour. 33, 445-452.

[39] L.W.McKenzie (1960) Matrices with dominant diagonals and economic theorie, Mathematical Methods in Social Sciences, Stanford University Press.

[40] N.Megiddo and M.Kojima (1977) On the existence and uniqueness of solutions in non-linear complimentarity theory, Math. Programming 12, 110-130.

[41] J.Milnor (1965) Topology from the differentiable view point, The University Press of Virginia, Charlottesville, Virginia.

[42] J.More and W.C.Rheinboldt (1973) On P- and S- functions and related classes of n-dimensional non-linear mappings, Linear Alg. and its Appl. 6, 45-68.

[43] K.G.Murty (1971) On a characterization of P-matrices SIAM J on Appl. Math. 20, 378-383.

[44] F.Nikaido (1968) Convex structures and Economic Theory, Academic Press, New York and London.

[45] H.Nikaido (1972) Relative shares and factor price equalization, Jour. of Intl. Economics 2, 257-264.

[46] H.Nikaido (1975) Economic adjustments under non-competetive pricing, adaptive Economic Models, Academic Press, New York.

[47] C.Olech (1963) On the global stability of an autonomous system in the plane, Contr. diff. equations 1, 389-400.

[48] J.M.Ortega and W.C.Rheinboldt (1970) Iterative solution of non-linear equations in several variables, Academic Press, New York.

[49] G.Owen (1968) Game Theory, W.B.Saunders Company, Philadelphia.

[50] R.S.Palais (1959) Natural operations on differential forms, Trans. Amer. Math. Soc. 92, 125-141.

[51] S.R.Paranjape (1978) Simpler proofs for the infinite divisibility of multivariate gamma distributions, Sankhyā Series A 40, 393-398.

[52] T.Parthasarathy (1982) On the global stability of an autonomous system on the plane, Technical report No. 8203, Indian Statistical Institute, New Delhi [submitted for publication].

[53] T.Parthasarathy, On univalent mappings (under preparation).

[54] T.Parthasarathy and T.E.S.Raghavan (1971) Some topics in two-person games, American Elsevier Company, New York.

[55] R.Plastock (1974) Homeomorphisms between Banach spaces, Trans. Amer. Math. Soc. 200, 169-183.

[56] R.J.Plemmons (1977) M-Matrix characterizations I-Nonsingular M-matrices, Lin. Alg. And its Appl. 18, 175-188.

[57] B.H.Pourciau (1977) Analysis and Optimization of Lipschitz Continuous Mappings, Jour. Opt. Theory and its Appl. 22, 311-351.

[58] B.H.Pourciau (1982) Hadamard's theorem for locally Lipschitzian maps, Jour. Math. Analy Appl. 85, 279-285.

[59] Rado-Reichelderfer (1955) Continuous transformations in analysis, Springer-Verlag.

[60] M.Radulescu and S.Radulescu (1980) Global inversion theorems and applications to differential equations, Non-linear Analysis, Theory, Methods and Applications 4, 951-965.

[61] T.E.S. Raghavan (1978) Completely mixed games and M-Matrices, Lin.Alg. And its Appl. 21, 35-45.

[62] P.S.S.N.V.P. Rao (1977) A characterization of merely positive subdefinite matrices and related results, Sankhyā Series A, 39, 387-395.

[63] Rothe (1976) Expository introduction to some aspects of degree theory, published in "Nonlinear function analysis and differential equations" (edited by Cesari, Kannan, Schuur), Marcel Dekker, Lectures in pure and applied mathematics No. 19.

[64] P.A.Samuelson (1953) Prices of factors and goods in general equilibrium, Rev. Econ. Stud. 21, 1-20.

[65] Collected Scientific papers of Paul A.Samuelson, Vol. II edited by J.E. Stiglitz (1965), The M.I.T. Press, Cambridge, Massachusetts pp. 902-908.

[66] H.Samuelson,R.M.Thrall and O.Wesler (1958) A partition theorem of euclidean n-space, Proc. Amer. Math. Soc. 9, 805-807.

[67] A.Sard (1942) The measure of critical values of differentiable maps, Bull. Amer. Math. Soc. 48, 883-890.

[68] R.Schramm (1978) Boundary conditions for the univalence of NVL-transformations, Math. Ann. 236, 191-198.

[69] R.Schramm (1979) Bounds for the number of solutions of certain piecewise linear equations, Proceedings of a conference on Geometry and Differential Geometry, Haiffa University, Springer Verlag, Heidelberg, New York.

[70] R.Schramm (1980) On piecewise linear functions and piecewise linear equations, Math. O.R. 5, 510-522.

[71] G.Vodossich (1980) Two remarks on the stability of ordinary differential equations, Non-linear Analysis, Theory, Methods and Appl. 4, 967-974.

[72] G.T.Whyburn (1942) Analytic Topology, Amer. Math. Soc.

[73] G.T.Whyburn (1958) Topological Analysis, Princeton University Press.

[74] G.T.Whyburn (1951) An open mapping approach to Hurwitz's theorem, Trans. Amer. Math. Soc. 71, 113-119.

[75] G.T.Whyburn (1961) Open mappings on 2-dimensional manifolds, Jour. Math. and Mech. 10, 181-198.

INDEX